T0164578

Never Summer

A Novel

Mark H. Gaffney

NEVER SUMMER: A NOVEL

Published by:
Trine Day LLC
PO Box 577
Walterville, OR 97489
1-800-556-2012
www.TrineDay.com
publisher@TrineDay.net

Library of Congress Control Number: 2016958657

Gaffney, Mark —1st ed.
p. cm.
Includes references and index.
Epub (ISBN-13) 978-1-63424-130-4
Mobi (ISBN-13) 978-1-63424-131-1
Print (ISBN-13) 978-1-63424-129-8
1. Fiction I. Gaffney, Mark H. II. Title

FIRST EDITION
10 9 8 7 6 5 4 3 2 1

Printed in the USA
Distribution to the Trade by:
Independent Publishers Group (IPG)
814 North Franklin Street
Chicago, Illinois 60610
312.337.0747
www.ipgbook.com

to Jeanie

The best time to plant a tree was twenty years ago. The next best time is now.

– Chinese proverb

Orlando:

Hang there, my verse, in witness of my love:
And thou, thrice-crowned queen of night, survey
With thy chaste eye, from thy pale sphere above,
Thy huntress' name, that my full life doth sway.
O Rosalind! these trees shall be my books,
And in their barks my thoughts I'll character,
That every eye, which in this forest looks,
Shall see thy virtue witness'd everywhere.

– William Shalespeare,
As You Like It, Act III. Scene II

It has been said that trees are imperfect men, and seem to bemoan their imprisonment rooted in the ground. But they never seem so to me. I never saw a discontented tree. They grip the ground as though they liked it, and though fast rooted they travel about as far as we do. They go wandering forth in all directions with every wind, going and coming like ourselves, traveling with us around the sun two million miles a day, and through space heaven knows how fast and far!

– John Muir, July 1890

BOOK ONE

ONE

The late afternoon sun was ablaze on the logging deck, winking off bright steel surfaces polished by long use, the wages of sweat and diesel.

The enormous clear-cut surrounding the dock was a scene of carnage – raw stumps and hip-deep slash, the remnants of an ancient forest whose time had come. The air, by contrast, was sweetly fragrant with conifer, death's lingering afterbirth.

At one side loomed an enormous mountain of recently skidded logs, ready to be loaded up and hauled away, next morning.

A few chainsaws were still screaming in the woods, men laboring for their keep. But most of the crew had called it quits for the day.

A half dozen men in work clothes milled about on the deck. The mood was laid-back, with a hint of tired bones. Two loggers sipped coffee from a thermos as they harangued a third man about nothing in particular, the ribbing all in jest but with just enough edge to push their comrade across the threshold of a smile. When his stony demeanor broke to the humor, the other two exchanged a glance. One sealed the moment with a raucous laugh.

Crew boss Jacques St. Clair climbed down off his D-6 Caterpillar and stood among his men. The chief removed his leather gloves and casually slapped his leg as he eyed his fifteen-ton tractor with a mix of pride and affection. The word "Puss" was inscribed in black letters under a big-busted nude on the yellow chassis. The cat's six-cylinder 120-horsepower diesel engine rumbled smoothly at idle. A hot column of blue exhaust rose from the stack.

St. Clair had a hole in his chiseled face. It flickered darkly when he talked like the negative of a flashing sign. The hole was no larger than the previous occupant, a front incisor. But everything is relative and the absent enamel grabbed the eye like no ivory ever could. Most of the time, when the boss spoke, the dark hole merely teased. But when his blood rose as it often did, the empty space loomed front and center, larger than life.

Nearby stood a scruffy-haired kid named Tom Lacey. Tom was skinny as a bean pole and wore wire-frame glasses. He was the smallest man on the crew by at least forty pounds, and the shortest by several inches. The brainy sort, Tom could not help the fact that he was inquisitive by nature. On his first day on the job, the previous spring, when Jacques had explained the ropes, Tom wanted to ask point-blank about that missing front tooth. "Boss, don't mean to be rude but how did you come by that hole in your mouth? Mischance? An accident maybe?"

The answer had come in time. Now, seasoned by a year running a chainsaw, Tom understood that all manner of bad things happen to men who toil in the woods for a living. It is a dangerous profession.

As the loggers turned for home and strolled toward the trucks, saws slung over their shoulders, the long bars tipped down, balanced easily by gloved hands, a six-foot-four monster of a man named Jimmy Thurston came shagging up to the deck. The big logger was empty-handed, without his chainsaw, and was mumbling under his breath. The man looked defeated, diminished despite his great size. Thurston was an irresolute sort of man, a bit dim around the edges, which accounted for his nickname, "Fuzzy."

"Where's your saw?" Jacques wanted to know.

"Boss, I fucked up," said Thurston, motioning with his head.

Someone said, "Look at *that* leaner! Holy shit!"

Overhearing, the other men stopped and turned.

"Good Christ!"

"A widow-maker!"

The problem was clear at a glance. Thurston had been working one of the last patches of standing timber, but in the course of dropping a big fir had made his back cut too low. The tree toppled the wrong way. In logger parlance, it "came back on him." The cut-through trunk was hung-up in a nearby crown. Thurston had then dropped a larger tree to shake the hanger free and take it down. But the second tree also got hung up, compounding the problem. Ditto on his third attempt. The fourth cut made things worse yet, as Thurston had jammed his saw beyond redemption.

Thurston now had five trees tangled up together, totaling several tons of biomass. In addition to that, his chainsaw was out of action. There was no easy way to drop the mess without grave risk. When cut-through stems are tangled in this way it's nearly impossible to know how they will go down.

It was a logger's nightmare. Even a glancing blow from an errant trunk can kill a man in an instant. Believe it.

Thurston was unnerved and had walked away from it. Now, he began to importune Jacques. He wanted the boss to solve his problem, even at the risk of damaging or destroying his saw, by pushing the tangled mess over with the D-6 cat. But the chief would have none of it.

"No way. Borrow a saw and clean up your own mess."

A shadow crossed Jimmy's face. The man shifted his feet and sunk his hands deeper in his pockets. No solution was coming from that direction.

The boss set his jaw. There was a long unpleasant silence on the landing.

"Hey, not a problem," said Tom breezily. "I'll take care of it."

"No, kid," said Jacques, reaching out. "Wait! It's too dangerous."

But Tom was already striding toward the mess, saw in hand. He scampered over and through the green trash like he was walking on water.

The landing fell silent. All eyes were watching.

Moving with fluid ease amidst the tangled trunks, Tom studied it up and down for maybe twenty seconds. Then, he started his saw with a yank and made a single cut. Wood chips began to fly. Moments later, there was a loud *snaaaapping* and a thunderous explosion as it all came crashing down.

There was an audible gasp from the men on the deck as Tom disappeared in a cloud of dust and flying fir needles.

"Oh shit!" someone said.

But as the dust settled, Tom reappeared holding up Thurston's chainsaw like it was a trophy. A shout went up from the landing as the men cheered him. One guy even threw his tin hat in the air.

"Way to go, kid!"

"All right!"

"Hallelujah!"

"Nice work."

TWO

Tom Lacey derived a strange and ecstatic joy from running a chainsaw. It was a hands-on kind of thrill and satisfied something in him, something that he had missed (without ever knowing it) during his failed time at the university. That previous period had been an inglorious abortion. He was happy it was behind him and to have moved on.

But if Tom was a refugee from his past, he gave little indication.

At home in the woods, he liked the gritty work. Not that he meant to stick with it; he had no plan to make a career of logging. Truth was, he had no plans for much of anything, having given up thinking about the future or what tomorrow might bring. He had fallen into a pattern of simply living day to day, just taking things as they came. The work was fine for now and that was good enough. Let the future take care of itself.

Running a chainsaw was all about the existential present, being in the visceral now. It required a high degree of situational awareness. A man who let his mind wander off while running a chainsaw risked serious injury or worse in the blink-of-an-eye. There was almost no margin for error. This was the weird flip-side, the hallelujah part of it, because same the same element of risk that kept a man on his toes also made the work interesting. Without the risk it would have been ordinary, humdrum.

Tom lived for the surge, the irresistible bite of the sharp chain, thc power of the saw working its will upon the wood. From the moment the old man had first thrust that beat-out Homelite into his hands, more than a year before, he had taken to it with an alacrity that astounded him (and the old man to boot).

Though after a year in the big timber Tom was now a seasoned cutter, the first weeks on Jacques crew had been rough, mainly due to his small size. Not that there was any doubt that the skinny kid with the granny glasses had an aptitude for teasing performance out of a chainsaw.

The boss certainly had been impressed. Jacques liked the way his new man "got after it." Tom did not just work; he attacked the timber. The kid did not know the meaning of "lazy" or "slouch" or "take it easy." In Jacques' experience this was unusual. It occurred to him that the kid might be working off a pretty big chip on his shoulder. But hey, so what? Some of the toughest guys Jacques had ever known were small men who compensated for it with grit.

Yet, despite Tom's aptitude for the work, initially most of the crew had been wary of him. They kept their distance, probably because they did not know what to make of him. Tom was different.

"He's a philosophy student for Crissake," they objected, but Jacques just laughed at them. When this had no effect they complained, feigning indignation, that he was too damn small to be knocking down large stems; as if they were affronted by the fact. True enough, Tom stood only five-feet-six (barely) and weighed a hundred and twenty pounds (in his boots). Next to them he was a runt, a dwarf among giants. This in a trade dominated by brutes since the gyppo days and steam donkeys.

But Jacques scoffed at them. What did size have to do with it? What did size have to do with anything? Completely irrelevant. The bottom line was production, pure and simple. The boss knew a man didn't have to be Paul Bunyan anymore to make it in the woods. So what if the trade had descended from stringy six-foot men capable of working a misery whip (a two-man cross-cut saw) all day? Things had changed and greatly since the days of the springboard and double-headed axe. All of that was now ancient history thanks to gas motors, miniaturization, lightweight alloys and space age plastics.

With the advent of the new lightweight saws, muscle had given way. Armed with one of the late models an average sized Joe, even a small man, could cut circles around macho loggers hamstrung by outdated equipment. It was a paradox that even as the labor force shrank over the years, the woods had opened up to smaller men, yes, even runts like Tom Lacey. It was why Jacques was willing to give the kid a shot. "Hell, I'd do the same for any man who wants to work. Besides, I need cutters. I'm short handed." Jacques was always short on loggers.

It certainly did not take him long to size up a new prospect. Shove an idling chainsaw into a man's hands and an experienced eye can tell within a matter of minutes if he has what it takes. Bingo! In the boss's experience most guys flunk the test. Even big men. Why? Simple. Most guys flinch at the compact fury of a chainsaw. They are intimidated, by the oh-so-short leash, by the ferocious proximity of all of that tightly-bundled power. Many men are cowered by the decibels alone, driven to distraction by the insane racket, by engine noise so earsplitting a man cannot hear himself think. It's why loggers must learn to "feel" their surroundings.

Not to mention the chain. The menace of three-dozen razor sharp teeth just inches from a man's leg can be hair-raising. Imagine the indifference of a machete slicing through the soft flesh of a watermelon and you will appreciate the extraordinary vulnerability of a chainsaw operator, shielded from the swift shredding of human flesh by skill alone.

It's why the average Joe hesitates, pauses to think about it; a fatal weakness in a timber faller. No wonder most men soon dispense with "that crazy idea" of running a chainsaw professionally, for safer and saner pursuits. As they say: "It ain't for everyone."

But not Tom Lacey. The kid did not know the meaning of hesitation. As Jacques observed with wry approval, with-saw-in-hand Tom was like a well-oiled machine. He never took a break except to take a leak, or to gas up his Husky. He had only one gear. Full out. Open throttle. The kid was a natural born timber faller, no question about it.

Eventually, Jacque became annoyed with the crew's bullshit. "You guys, I'm sick of your whining. Like a bunch of old women. Sure the kid's green, but he's already out-producing most of you. He stays, and if you don't like it, sue me. Better yet, kiss my ass."

THREE

Jacques' judgment was spot on. Before the first week was done, Tom had demonstrated to all concerned that he could handle a saw with the best of them. He was a natural, a virtuoso. The men shut off their saws just to watch him work.

"What's with that guy?"

"He's one eager beaver, ain't he?"

"Look at the way he drives that thing."

"Shit. Gangbusters."

"He's scrappy. I'll give him that."

"That son-of-a-bitch works like there's no tomorrow."

Before long, the ribbing had turned respectful. The men warmed to him. One logger joked that Tom's skin was so pale because he had ice water in his veins.

What a motley bunch they were. Tom soon discovered that some of the men, "Red" Callahan, a veteran timber faller, hard-boiled by eighteen years in the woods, Charlie McCoy, and "Dipstick" Dugan, were regular fellows that anyone would be proud to call a friend. But he also learned to steer clear of a few less savory individuals. One logger known simply as "the Preacher" gave the appearance of being a hillbilly and more than lived up to it. The man wore baggy pants held up by suspenders, and crummy boots that might have passed for clod-hoppers. He had a long beard that he liked to fondle and beady little eyes that gawked from under his bushy brow. The Preacher never missed an opportunity to talk about his favorite subject, his own personal salvation; hence, the nickname. Every member of the crew had been subjected on multiple occasions to his long-winded religious harangues. Worse, the Preacher had a tiresome habit of wagging his finger at you, usually in your face, when things did not meet his fastidious approval. When the crew laughed at him, as they invariably did, it only fed his born-again fervor.

Wolfe Withers was another case. The man had recently been released from Canon City where, according to word around camp, he had done time for murder. Not even Jacques, though, knew the details. Wolfe had a swarthy complexion and a menacing attitude. An air of dark mystery surrounded him because he was a loner, completely asocial, and made no effort to dispel the rumors. When the men tried to engage him in friendly conversation they encountered his nasty disposition. One or two attempts was enough. After that, they backed off and gave him a wide berth.

And then there was Shorty Dibbs...

The next Friday was payday, and about quitting time the boss showed up on the landing to pass out the checks. The loggers gathered around.

The married men usually drove home to be with their families on weekends. On payday, though, they would first make a beeline to the bank in Granby to cash or deposit their checks, then, scattered in all directions.

On this day Red Callahan and Charlie McCoy strolled across the street to have a cold one at the Nugget before leaving, and invited Tom and Shorty to join them. The saloon was packed and rowdy on a typical Friday afternoon. The Coors was flowing freely.

The loggers lined up at the bar and Shorty insisted on buying the first round. He was 200 plus pounds and stood well over six feet tall, but was no mental giant, hence, the nickname. Tom wondered why Shorty never took offense when they called him that, but he was beginning to understand. The man was witless as a snowflake.

After one beer Red and Charlie left for home but Shorty was just getting started. The man was soft-spoken and had a gentle disposition, basically good company so long as he was dry. However, alcohol affected him powerfully, and not for the better. Tom watched as Shorty tossed down double shots of Jack Daniels, back to back, then, called for another. The big logger was on his way to being sloppy drunk. Tom was still working on his first beer when Shorty turned and began talking to a fellow Tom did not know.

"Hey, Jack, had any lately."

"All the time," the other man said right back. "An' it still ain't enough." Both guffawed.

"Here's to ya." Shorty was now slurring his words.

"Down the hatch," said the man who then took a turn. "Shorty, did I tell you the one about the Rocky Mountain oysters?" Before Shorty could say a word he was off to the races.

Tom was watching the people at the bar, relaxing as he listened.

"Had a buddy who told it to me after he got back from Mexico. Y'now, they love their bullfights down there. So, well, my friend was in a cantina sipping his sangria when he noticed a waiter serving up a dish at the next table that looked and smelled so good he asked the waiter what it was. The man in the white jacket told him it was a rare delicacy, deep fried bull testicles. Okay, my friend says, 'Well, it looks so good I think I'll have the same myself.' 'Sorry, señor,' the waiter told him, 'supplies are limited and we are out. But if you come back the day after tomorrow, after the next bullfight, I'll be happy to save some for you. Por seguro.' Well, two days later my buddy went back and placed his order. Sure enough, the oysters were delicious, as described. But they were a lot smaller than the ones on the previous occasion. When he asked the waiter 'Why is that?' the man just shrugged and said, 'So sorry, señor, but...how you say...algunas veces?...aah...some times...the bull he wins. Ha!" The guy howled so hard at his own joke he slipped off the barstool. But he climbed back up and clapped Shorty on the shoulder like a good old boy. Tom missed part of it, but got the punch line. But Shorty never did. He just stared at his drink.

That's when the trouble started. A green-eyed stranger had stepped up behind Shorty, and now grabbed him by the shoulder, swung him around and got right in his face. Evidently the man had taken offense at the *Timber Forever!* logo stenciled on the back of Shorty's T-shirt.

"I know what you do," the stranger said in an outraged tone. "I've seen enough of your clearcuts to last me a life time." It was too weird. The stranger began to lecture Shorty about leaving something for future generations; a mistake, as it turned out.

"Pshaw. What did future generations ever do for me?" Shorty said as he set down his drink and leaned back, elbows on the bar. The slur was gone now and his voice had an edge that Tom found alarming. Shorty's body language was an explicit warning, one the other man would have been smart to heed. But apparently he failed to notice or just did not care.

Tom found it hard to believe his own eyes. Was this really happening?

The stranger put it right out there. He began to lecture Shorty about a place called Bowen Gulch that he said was "God's country." He and his friends were going to save it no matter what it took, come hell or high water.

"Happy to oblige," said Shorty and slammed the man's head against the bar. The body went limp and slithered to the floor. Shorty was on him fast, then, kicking and stomping until Tom and another man pulled him off, probably the only thing that saved the green-eyed stranger's life.

As Tom attended to the fallen man, Shorty shambled out the side door to piss in the alley under the lonesome Colorado stars. He was back to slurring.

FOUR

Working in the woods was much the same from day to day. The logging routine hardly varied. But Tom had no issue with the monotony. He was having so much fun running his chainsaw that he could hardly wait to get out there and do it again. He was like a kid with a shiny new toy.

A summer high had established itself over the Rockies. The weather continued clear and mild, though in the first week of July the night temperatures dipped sharply, not unusual in the Rockies.

On the morning of the last hard frost of the season Tom was out of the sack before anyone else. He was already enjoying his first steaming cup of coffee when the logging camp began to stir.

Tom had the jump on the day. He ate in haste and was still working on the last of his eggs when he fired up his rig. As the engine warmed up he made small talk with Charlie McCoy, another early bird. Charlie was loading gear into the back of a truck. The older man yawned and rubbed a fist in one eye. "Brrr, it's cold."

"Want some coffee?" said Tom. "There's more."

"No thanks, I'm coffee'd out." Charlie closed the tailgate and looked at the sky. "Looks like another good one. Later, kid."

Minutes later, Tom reached the logging site and parked his pickup just above the main landing, grabbed his gear, and started into the timber. He tramped down-slope through dense coniferous forest, then left the stand and started across a wide right-of-way clear cut, soon to be the site of a new reservoir. The stump field was heavy with frost and still deep in shadow.

Hardly noticing the devastation, Tom paused to savor the morning. High above, the mountain ridge was aflame, brightly backlit by sun. The kid shivered. No matter. The chill would pass in a hurry once the sun broke over the feathered rim.

The best time of day!

The footing was bad as he picked his way over the frozen ground, then, reentered the forest and followed the yellow flagging to his allotment.

Deep in the stand he stopped and unslung the two plastic jugs joined by a nylon cord from around his neck. One held oil for his saw chain, the other his gas mix. Setting the jugs aside, he took his Husky in hand, flipped the "ON" switch and set the choke. With one hand on the handle to steady it, he gave the rip-cord a smart yank. On the third pull the saw popped and sputtered. He released the choke and gave it another sharp pull. The saw was now primed and came snarling to life, fuming steel-blue smoke. Its high-pitched wail and the haunting echo, up and down the valley, shattered the peace of the mountain morning.

Tom was used to the racket but grimaced anyway. He looked around as he revved the saw. Here there was no "edge" to work. He was starting deep in the interior of the stand. His eye settled on his first tree of the day, a smooth-barked subalpine fir, about two feet in diameter. He eyed it up and down. Throttling the saw, he cleared the lower trunk of small branches with several quick vertical swipes.

Without further ado he moved into position and went to work. How he loved that first bite, the surging saw in his hands.

He worked easily, without strain, relaxed within himself as he guided the chain through the tree. The trick was to let the saw do the work. The teeth were razor sharp. The chainsaw was made to devour conifer and, tapped out to the max, the four cubic-inch two-cycle engine ate wood as fast as he could feed it. As the cut deepened large curlicue chips spewed out in a pile at his feet. The aromatic smell of pitch filled the air, sweet as a kiss, fresh as the morning air.

It doesn't get any better than this.

The feeling of so much power at his command was indescribable. It was the best part of the work. The tree was putty in his hands.

A little more than halfway through the trunk he backed out and made his second cut, above and at a downward angle to the first. Then, he swung the bar out. As he did, a small wedge of wood blew out in a flurry of sawdust and shavings.

Deftly shifting position, he moved around the tree and started the back-cut, the *coup de grace,* slightly above the frontal cut. Halfway through, he paused and tapped a plastic wedge in behind the blade using a small axe as a hammer. The wedge was insurance. It would prevent the big fir from coming back on him.

He throttled the saw again to the max, every sense awake to the slight shudder he knew so well, the feel of tree beginning to give way. Gravity at work. The feeling was exquisite and when it came he pulled out fast and stepped back, out of harm's way.

Only then did he look up.

The topple started slowly. From the look of the crown he knew it was a beauty. The fir was going to lay down exactly where he wanted it. Slowly it gathered momentum; then, came the thunderous crash with pine needles, bark, and branches flying.

A deafening *KA-WOOOMPH* momentarily drowned out his saw.

His head was a void as he trimmed the butt to even it up. He liked to stay in a thought free zone – a kind of no-mind Zen headspace.

He worked his way horizontally up the trunk toward the crown, methodically lopping branches off the fallen giant. One last cut topped the bole. Done. His first tree of the day was now a log. He looked up, searching for the next one.

That one.

He strode toward another large fir and repeated the process. Before long, three neatly trimmed logs were on the ground ready to be winched up with choker cables and skidded to the loading dock. There, a bucker would measure and cut each one to length.

Three trees down and he had his opening, a small hole in the canopy where he could continue dropping timber without undue concern about hangers and wayward crowns.

By 9:00 A.M. he had peeled down to his t-shirt and was working up a sweat. But now the saw surged, the sound of a too-lean mix. The Husky was low on gas. He shut it off, but the high-pitched wail continued in his head. Layered over this, in the background, was the sing-song medley, rising and falling, of other chainsaws screaming in the distance, the sound of St. Clair's loggers at work, a dozen men getting after it.

Tom looked up and noticed he had an audience. Dipstick and Shorty waved from across the way. The two had been watching him work. By now, this was almost commonplace. He waved back.

"Leave some for us, will ya," Dipstick shouted.

"Yeah, kid, don't be so greedy," said Shorty. Smiling, Tom joked with them and exchanged pleasantries, until the men finally picked up their saws and went back to work.

He leisurely attended to his Husky. First, he replenished the gas and bar oil, then, touched up the chain with a Swedish round file. From the draggy feel of the bar he knew he had dulled some teeth.

As he was filing he lost his concentration and his mind wandered. Suddenly, he was thinking about her. Tallie. A waking dream real as life. The previous winter, their torrid affair in the Florida outback had ended when he reluctantly put her on a bus to California. Months later, after his subsequent letters went unreturned, he had decided to expunge her forever from his thoughts; and had succeeded until her letter arrived via General Delivery. As much as he wanted to read it, he nonetheless had stashed it under his bedroll, heartsick, dreading what it might contain. The unopened letter had plunged him back into the maelstrom. He was now shocked by his own thoughts. It occurred to him that the work had become his drug of choice. Never had he imagined that even mindfulness could be a means of escape...

Pushing such thoughts aside, however, he was soon back at it, dropping trees, which is how the afternoon passed, just another day in the woods.

The sun was still high in the west when he gathered up his gear and headed for camp.

FIVE

Pinecone Peters slammed down the receiver. He was pissed beyond words. He'd been working the phone all afternoon, indeed, most of the week, and had nothing to show for it. He had just spoken with the lead attorney at the Colorado Environmental Coalition. The man had taken twenty minutes to explain in mind-numbing detail why the CEC had decided *not* to pursue a lawsuit to halt the Bowen Gulch timber sale.

"It's too late," the attorney told him. "As you know we missed the deadline for the administrative appeal and we think our chance of success at this point in the courts is not very good. A judge would probably rule that we have no standing. Our staff and resources at CEC are limited as I'm sure you know. So...we've decided to focus our energies where we think we can have a real impact. We...uh, just feel that we have a better shot with some other projects. I'm sorry about this. I know how you feel about Bowen Gulch. We feel the same way but we're just slammed. If we had a larger staff, maybe...It was a tough call for us."

Everything the attorney said made perfect sense. Pinecone could not fault the logic, but dammit it still sounded like defeatism. He had been hearing the same story all week. He had heard it from state and local environmental groups, and from the big nationals like Audubon, the Wilderness Society and the World Wildlife Fund. In a funk, he had even tried the Isaac Walton League and Trout Unlimited, with the same result. Nothing.

No one could explain exactly how it had happened, how the environmental activists whose job it was to watchdog the Forest Service somehow missed the deadline to appeal the decision for the Bowen Gulch timber sale. It was a collective woops, and in the wake of this monumental screw-up the conservation community had turned gun-shy. They had given up without a fight, thrown in the towel; everyone, that is, except for a radical fringe group named Earth First!, which had already staged several demonstrations to protest the sale. Earth First! had become notorious for direct actions like tree-sitting, tree-spiking and monkey-wrenching, "eco-defense" they called it. Pinecone had friends in the group and occasionally joined them in direct actions. He found himself gravitating toward their radical viewpoint because in his view the Earth Firsters had their priorities straight, one of the few environmental groups that did. Earth First! believed that positive social change never happens without jeopardy. In other words, society only changes for the better when courageous individuals put their asses on the line for something they believe is worth fighting for, and if necessary, dying for. Needless to say, their risky style was not for the faint of heart.

Pinecone also knew that direct action by Earth First! alone would not succeed in halting this particular sale. Numbers mattered. Bowen Gulch would only be saved if the wider public could somehow be induced to weigh in.

He gently massaged the base of his nose. His face throbbed and itched at the same time. The body of the nose was bulbous, still swollen and tender from the thrashing that deranged logger had given him in the Nugget the previous week.

Pinecone was frustrated, yes, but even more than that he was disgusted. The folks who ought to be acting were doing nothing to prevent the despoliation of one of North America's last best places. If words still had any meaning, Bowen Gulch surely qualified as 'special.' Located in the Never Summer Mountain Range near the western boundary of Rocky Mountain National Park, the high valley featured the most impressive stand of Engelmann spruce in the southern Rockies.

Pinecone knew the place well. He had walked the Gulch many times while employed by the US Forest Service. An afternoon spent wandering through the high forest always rejuvenated him. The Gulch was special not only because of the remarkable size of the trees – some of the spruces measured five feet in diameter – but also because of their antiquity. A few weeks before, he had consulted a retired forester who told him that many of the big trees were in excess of 800 years old. The old ranger had cored numerous trees in Bowen Gulch and by counting the rings had determined their age. The place deserved protection solely on this basis. The rare combination of antiquity and structural decadence made Bowen Gulch a biodiversity stronghold. The place was a haven for all manner of wildlife, including several rare and endangered species.

The problem, of course, was that the Forest Service did not see it this way. The timber beasts who ran the agency regarded great age and structural decadence as a liability, not an asset. "Over mature," they called it. It was one of the terms they bandied about that had convinced Pinecone that ranger Dougie Bennett and his timber sale staffers were out of their minds. They had it exactly backwards. Their "cure" was the problem. The current policy of extraction meant replacing ancient forest with "young thrifty stands," another mad expression. The problem was that most of the old growth was already gone. A recent inventory had shown that after more than a century of logging less than 10% of national forest land remained in a pristine state. From the standpoint of biodiversity, the numbers were grim.

Convinced that the best strategy to save Bowen Gulch was to share it with others, the previous summer Pinecone had begun leading hikes into the area, often on Sundays. Hiking the Gulch was like being in church, only better. The cathedral forests along Bowen creek invariably reduced visitors to respectful silence, even reverence. The place was the real deal. Seeing was believing. A single visit usually did the trick. People came away passionate about saving it. Many were outraged that the agency would even consider logging such a place. Unfortunately, building this kind of awareness was a slow process, and time was short.

For weeks Pinecone had been searching for someone to lead the campaign to stop the sale. It had to be someone of stature who was well connected, and already known to the broader community. The "someone" had to be tough enough to take the heat and charismatic enough to rally others. But that someone was proving elusive. After weeks of searching Pinecone had come up dry. Was he chasing a ghost?

Starting from the top, he went down the list one more time. He had already contacted everyone on it at least once. Each name had a check mark beside it.

Wait a minute. Ah, yes...

All, that is, save for one. He had forgotten. One name was still unchecked, Dr. Mickey Newsome, a math professor at the University of Colorado who for many years had chaired the Boulder chapter of the Sierra Club. For weeks Pinecone had been trying to reach Newsome, without success. He had called him a dozen times but apparently the man was out of town.

What the heck. Nothing to lose at this point.

He picked up the phone, dialed and listened for the ring.

"Hello."

"Dr. Newsome?"

"Speaking. Who's this?"

"Sir, you probably don't remember me. We met last year at a Sierra Club event. My name is Pinecone Peters."

"Call me Mickey. It was a potluck and I do remember you. I never forget a face, or a voice."

"I've been trying to reach you for weeks."

"I was out of town."

"That's what your wife told me."

"I'm just back from South America. I'm a climber, you know."

"Really? Sounds like fun."

"Yeah, dangerous as hell but fun too, you bet." There was a chuckle on the other end of the line. "So, what can I do for you?"

Pinecone told him about the decision to log Bowen Gulch.

"Damn. I know the place."

"Then you know how special it is."

"I thought Bowen Gulch was still under roadless area review."

"No, they finished that almost a year ago."

"Hmmm. This is not good."

"The day after the roadless review staff at the regional office released Bowen Gulch for timber production, the local ranger at the district office assembled a project team and started planning the sale. They intend to go in as soon as possible and gut the place."

"How large is the sale?"

"It's big. Ten million board feet, and that's just for starters." He explained that the language of the sale called for multiple entries. Bowen Gulch would be re-logged in ten years and again after twenty. Pinecone heard a whistle over the line.

"How come nobody appealed it? What about the folks at the Colorado Environmental Coalition?"

"They slipped it by them. The CEC are spread thin like the rest of us. Too many things to stay on top of."

"Yeah, I know. Same old story. And too few people to cover the bases. Well, the bottom line is, we can't let them do it."

"I agree."

"Over my dead body."

"That's what I said and a guy broke my nose."

"What?"

He told Newsome about the incident at the Nugget.

"I'm sorry to hear about that." There was a long silence on the other end. "My sense is, we need to jump on this. Right away. Where are you based?"

"Granby."

"That could be helpful."

"What are you thinking?"

"We need to set up a meeting with the ranger who's pushing the sale."

"That's my former boss. 'Dougie' Bennett. I know him well."

"So, you know him. What do you think the odds are of persuading ranger Bennett to drop it?"

"Try a snowball in hell. He's a timber beast. I ought to know. I worked with him for three years."

"Then, it's gonna be a fight."

"Yes."

"OK. But we still have to observe the formalities. The sooner we pay your old boss a visit the better."

SIX

Tom fixed his supper alone, marveling at the slow pageant of shifting light. The day was winding down, the sun in grand retreat.

She was in his thoughts.

He ate in silence watching the mountain valley compose itself for night. Slowly the shadows lengthened as the veil of dusk settled over camp. But the light show continued higher up, along the rocky ridge that jackknifed skyward above camp. As dusk slowly scaled the heights, the grand finale played out all the more brilliantly in rosy reds, lavenders and, finally, in shades of pink that lingered on the highest point, then faded into night.

From his site near the edge of camp Tom could hear the men moving about in the gathering darkness, their laughter and idle talk.

There'll be a fire tonight.

He stepped inside his tent and pulled on a heavy cardigan; then, pausing, he removed Tallie's still unopened letter from the pillow, turned it over in his hands and ran his thumb along the top edge as if to confirm it was real. He snapped on his flashlight and for a moment spotlighted the return address. San Francisco. He lifted the scented letter to his nose and inhaled deeply. The smell of her was like ambrosia.

Intoxicated, he set the letter back on the pillow, grabbed an apple and headed for Red Callahan's.

By now, cavernous night had swallowed the logging camp. The bonfire was blazing up with garish effect. Illuminated by the flames, the broad surfaces of tents and campers stood out in sharp relief. Man's last line of defense against...what? Absolute nothingness.

A half-dozen loggers stood around the fire like refugees or maybe fugitives. The flickering faces were spectral masks, lit up from below, the eyes lost in shadow. The bony contortions were a parody of human features. The drift of small talk was subdued.

Dipstick Dugan lifted a hand and proffered the remains of a six-pack dangling from its plastic yoke. "We got more Bud here. Who wants one?"

"I do," croaked Kermit Johnson.

Dugan slipped a can from the plastic ring and passed it over. "Who else? One left." He shoved the beer at Tom who waved it off and took a bite of his apple.

Charlie McCoy stepped out of the shadows into the circle of light.

"What say, Charlie?"

"Somebody talking to me?"

"Hell yes somebody's talking to you," said Dipstick. "I'm talking to you.... You mangy piece of shit." A ripple of laughs moved around the fire. Charlie was stroking a three-day beard.

"McCoy, you're one ugly hairy fucker. Don't you think he's ugly?"

"Course he's ugly."

"One ugly fuck."

McCoy looked up. "Yah, Dip, almost as ugly as your baby sister." More laughs.

"Them's fightin' words mister."

"Uh-oh. Now you done it. You frosted him."

"Look out."

Dipstick had his dukes up. "See this? This here's your face." Dipstick slammed his knuckles into the palm of his other hand.

"Scary guy."

"Take that bug out of your ass," Charlie said, "and hand me a beer. OK?"

For a moment Dugan stared as if in disbelief. He detached the can from the plastic. "Hokay, boss," he said.

"Last one." But instead of passing the beer across he heaved it. Charlie was ready, though, and caught the missile inches from his face. He shook it hard and as he popped the cap he pointed it back at Dipstick. A spray of frothing beer arced across the fire. Dipstick ducked but could not escape a bath of foam. He stood shaking his arm while Charlie sniggered. The antics produced more laughs all around.

"Much obliged," Charlie said with a nonchalant wave of the frothing can.

Someone taunted Dip not to settle for seconds. But the horseplay had run its course. The laughter subsided. The mood turned somber. The only sound was the crackling fire. A voice said, "What happened to your boot there, Jimmy?"

"OOOWEE!"

"Let me see that. Son of a bitch."

"Fuzzy had a close one."

"Had a slip-up today," Thurston said as he lifted his muddy boot onto a rock by the fire. The men crowded around for a closer look. It was true. The toe of the leather boot was gone, sliced clean away. Firelight danced on the wriggling toes.

"Will you look at that."

"Slick as a whistle."

"Jimmy likes to clip his toe nails with his Homelite," Charlie hooted.

"Yup, they say the sweet meat's close to the bone."

"Atta boy."

"SUU-ee! Bet he hasn't changed that sock in a week."

"Lucky for Thurston he's got retractable toes," another one chirped, amidst titters of laughter.

"You're righter than you know," Thurston said, laughing with the rest. "When it happened they tucked up tight and headed for cover. I didn't know for sure if I still had 'em until I looked."

"Ha. Jimmy got his GRE. He can count to five."

"That'll teach him respect."

"Laugh all you want, boys," said Sourpuss Malone with mild reprove. "But I guarantee you won't be laughing when it happens to *you.* And believe you me, loggers, it will happen. Matter of when, not if."

"Make way, there," Red Callahan said as he stepped into the circle with an armful of wood and chucked it onto the ground. Then, crouching on one knee he fed the blaze. A shower of sparks climbed into the opaque night.

"Ever had a close one, Sourpuss?"

"Does a bear shit in the woods?" he replied. Then, he added, "Sure, mon. Who hasn't?"

"More times than you want to remember, right Malone?"

"Got that right. Had one last year, in fact. One green logger dropped a big yellow pine that damn near took my head off. Coupla cunt hairs the other way an' I wouldn't be here telling ya." More laughs.

Now, the fresh fuel began to catch. The flames leaped higher and the circle of men brightened.

"The most dangerous thing in the woods," Charlie said, "is a goddamn green guy with a chainsaw. Green guys and saws don't mix. No way, no how."

"Ain't it the truth."

"Amen to that."

Someone passed Tom a pint of Johnny Walker. "Here, kid. Live a little." Tom took a slug and moved the bottle along.

"I dang near killed a fellow once, myself," Charlie said.

"Oh?"

"How's that, Charlie?"

"We was clearing timber along a road two three years back. As I recollect, a big right-of-way contract. They was going to widen the road. But it wasn't a major highway or nothing. It was that gravel road west of Red Feather Lakes, up in the boonies."

"Oh, I remember that job. On the North Fork of the Poudre River, right Charlie?"

"That's right. Kermit knows. He was there. 'Member how we didn't have no flaggers, Kerm?"

"Yep, I recall."

"Right. Because it was a back road and there wasn't very much traffic. Anyway, I was working the uphill side of the road, see? Just finishing my cut through a twenty-inch fir, when, guess what, along comes this Volkswagen bus out of nowhere tooling up the trail. Well Jesus H. Christ what could I do? I mean shit, it was too late to call a stop, and too late for damn sure to hold back that tree. So there she went, over and" – Charlie whistled and motioned with his arm – "FWACK! Down on the guy's windshield. I kid you not."

"Did you kill the guy, Charlie?"

"Yeah. What happened?"

"Lucky for him I was on the uphill side of the road. Because the high bank broke the fall. So, nothing much happened. Only a few top branches hit the car."

"What'd the guy do? Get out and stomp your ass?"

"You're kidding me, right? Hell no, that guy never stopped for nothing. He kept on trucking. The only thing that som' bitch stomped was the gas. Should've seen the way he dragged out of there. Ha-ha-ha. I mean, you could hear his foot hit the floor." Charlie paused amidst laughter. "Must of scared the living shit out of him. In fac', I know I did because I got a look as he was going by. The guy's face was white as a sheet, like he'd just seen the boogie man."

The men chuckled and passed the Johnny Walker.

"What about you, Tom?" someone said. "Ever had a close one?"

"Hell, he don't even have whiskers. Just peach fur."

"Yeah, did you?"

"Yes," Tom said. "Once."

"Only once?"

"Shut up, pea brain," said Red. "The man's a scholar. Let him speak."

"Tell us."

When Tom spoke he appeared larger than his five-six frame. "Yeah, I did have a close call, my third day on the job."

"Didn' you say you worked for that old-timer over in North Park. What's his name?"

"You mean Carl Olsen," Red said.

"That's right. Carl lined me out."

"Cantankerous old Swede."

"No, the old boy's actually Norwegian."

"He's into posts and poles, ain't he?"

"Yep."

"There's a world of difference."

"Oh," Sourpuss said. "I think I know who you mean. You talkin' about that old hunting guide from Walden who's been all over everywhere?"

"That's the one. Old battery acid."

"I heard about him. They say he's got a sixth sense."

"You heard right," Dipstick said. "One of my brothers hunted with him, once, few years back. Says the guy's a spook."

"Oh yeah?"

"He ain't guiding no more, though. The old boy packed it in."

"Dead?"

"No, no, retired. He don't hunt no more."

"Well, dead or alive, he's one salty fucker." More laughter. Then silence.

"But Tom was gawn't tell us..."

"Yeah, I want to hear 'bout that close one."

"I was in a stand of Engelmann Spruce," Tom said. "A kind of boggy place. You know how Engelmann likes wet ground."

"That's the one with the shaggy bark."

"Yup. Yup."

"See, what happened, I made some dumb-ass moves that nearly got me killed. It was late in the afternoon and I was tired. Dead on my feet. If I'd known better I would have called it a day. But I didn't have the sense to knock off. It was a big spruce. Must have been, I'd say, thirty inches across, breast height. OK, so I made my back cut too low, for starts. Then, for some reason I let my saw cut drift down at a bad angle. Oh and I also forgot to use my wedge."

"OOOOO," a voice said. "Not good."

"It was inadvertent, but..."

"What's that mean?" said Kermit.

"What's *what* mean?"

"That word he said, in-at-ver-dant?"

"It means he fucked up."

"Go on, Tom."

"Like I said, three wrong moves. So there I was, just finishing my back cut when that big spruce settled down on my bar. The tree was cut clean through. But I didn't know it yet. The trunk was just sitting there perched on that stump, balanced, maybe a ton of dead weight on the bar of my saw."

"Ho-lee fuck."

"There was no chance to squirm it free."

"Nope. Waste of time."

"Forget it."

"No way. She was stuck like the sword in the stone."

"No wind, eh?"

"Nope. No wind. If there'd been a breeze she'd have gone right over. But the woods were totally calm. No breeze at all. I didn't know that the stem was cut through. I..."

"Why didn't you wedge her over?"

"Yeah, what about your wedge?"

"Said he forgot."

"I was getting to it but I was moving too slow I guess. Before I could think what to do that big old spruce, uh, I never would have believed it if I hadn't seen it with my own eyes. She didn't fall away clean. Nosir. That spruce slipped off the stump, still vertical, and buried itself in the duff a couple inches from my boot."

"Lord A-mighty!"

"I stood there in shock. I couldn't believe it. Thing is, it slipped off without a waver and stayed vertical. It was only *after*, I'd say, another four or five seconds during which I watched my whole life pass before me that she finally began to tip a little. Then she went over. Now, I'm quick on my feet. But that slip happened so fast off that stump I never saw it coming…"

"Fuck-in-A. If it had landed on your boot it would have pinned you to the spot..."

"She would have come down right on you."

"Couple of inches t'other way…"

"You'd of been in a world of hurt."

"Crippled up."

"One dead monkey."

"Crushed like a bug."

"I got weak in the knees when I realized what'd happened. Had to sit down an' think about it..."

"Asshole puckered up too, I'll bet."

"Hah! Tom shit his pants!"

"Well, he ain't the first an' won't be the last."

"You was one lucky bastard."

"Dang, Tom, if you didn't have a close one."

Someone passed him the Johnny Walker. "Kill it, bub. You earned it."

SEVEN

Jacques St. Clair stepped up to have a word with the driver, then backed away and gave him a curt wave. The driver nodded. The big rig's powerful diesel engine *ga-hoomed*, pumping smoke out of both stacks as the driver inched the fully-loaded truck forward across the landing, barely serviceable now because of the mud. Fortunately, it was not deep enough to require chains. The driver paused briefly at the edge of the pavement, then, after checking both ways slowly pulled out onto the Poudre River highway, after which, he accelerated down the road, pipes snaking black soot.

Good bye and Amen.

Jacques glanced at his watch.

It was the sixth load out this fine July morning. More would follow in the afternoon. The drivers would return again in the morning and continue the process. Jacques was still playing catch-up, but the huge mountain of logs on the landing had visibly shrunk. Within another week, ten days at the most, they would haul away the last of it, and he would wrap up the job or, at any rate, his part in it. Already the state engineer and survey team were swarming over the high valley laying out the earthen-work dam to be constructed later in the summer. The headwaters of the Poudre River was an excellent site for water catchment in a state where agriculture is heavily depended on irrigation. The week before, the state boys had brought in the big earth movers and graders that would be used to construct the dam and shape the bottom of the planned reservoir. For the moment, the heavy equipment stood idle beside the highway. By then, of course, Jacques would have moved his operation across the Divide to the next big project, to log a place called Bowen Gulch.

The double whammy of the big wind followed by the early snow the previous fall that dumped three feet of the white stuff on the Rockies, forcing an early shutdown before Mike Garity was in the ground, even before Jacques could begin cleaning up the windfall mess, had put him in a foul temper all winter. Jacques had passed much of it consoling Anita, Mike's widow. It was hard living with the death of a man like Garity, a good friend who had worked for him for ten years. Jacques knew he himself was blameless; it was after all a cruel business. Still, he second-guessed himself constantly about it because he felt responsible. He couldn't help it. Nor was Garity his first employee killed on the job. Two years before, he lost a driver named Jerry who had never handled a chainsaw nor even set foot in the woods. The driver had arrived for a load of logs but had to wait his turn. The guy operating the cherry picker was still loading the truck ahead of him. Jerry paused to have a smoke and was casually bullshitting with the other driver when the operator dropped a log on his head. The man standing next to Jerry had not even been touched.

Go figure.

The worst part was encountering Garity's widow about town, and the two kids who would never see their father again. Jacques had been covering the family's rent to help out and did not mind a bit though he was under no legal obligation to do so; even though his own finances were strained at the moment because of that damn penalty clause in the contract. It had been costing him all winter, for every month the right-of-way project dragged on past the deadline. Stipulated deadlines were supposed to create an incentive. Such contracts usually involved a bonus. But Jacques thought the incentive system was a bunch of crap, just plain stupid. He did not need an incentive to work; he was self-motivated. He hated leaving projects unfinished about as much as he hated playing catch-up.

Due to the near-record snowpack and the unusually wet spring, the woods had not opened up until after the 4th of

July. His men were still slogging through deep mud. Fortunately, the job was nearly done. Another ten days and he would be caught up, and the next project would save his bacon. It was a high volume timber sale at a place called Bowen Gulch, near the south end of the Never Summer Range. The site was remote. Jacques had not yet walked it but had been told that most of the sale units included exceptionally large diameter old growth spruce. This was welcome news because it promised a substantial payday. The sale might even put him over the top and make his dream of early retirement come true sooner rather than later.

The boss looked at his watch, then waved to several of his loggers as they strolled by. With a sigh he glanced at his foreman, Francis Delacour, standing nearby. Whenever he thought about it, which was not often, Jacques knew he was fortunate to have in his employ a ramrod as able as Delacour. The man could tear down a skidder and reassemble it with his eyes closed. In addition to being a wizard mechanic and a talented heavy equipment operator, Francis was, thankfully, also soft-spoken and loyal without question. Best of all, he required no supervision. A word, a wave of an arm, sometimes just a smile or a nod, was all the direction the man ever needed.

Like Jacques, Francis was of French Canadian ancestry. Five years before, he had abandoned logged-over New England and come west in search of his fortune. St. Clair hired him on the spot and never once had cause to regret it.

Francis now stepped up to have a word with the boss. One of the skidders had a broken drive shaft and a decision had to be made about acquiring replacement parts as soon as possible. But when he saw the expression on Jacques' face he changed his mind and stepped back. It would have to wait. Part of his job was staying on the chief's good side, and that meant knowing when to speak up and when to hold his water.

St. Clair was not a difficult man to work for. Despite his moody ways and occasional white-hot temper, the boss played fair with his men. Most importantly, he paid on time. Jacques was honest and, as bosses went, one of the best. Francis had worked for far worse. But he also knew from experience that when the chief had that look it was best to stand back. Turning, he headed off in the direction of skidder number two. The chief would come around when he was ready to talk.

Jacques never noticed his foreman leave. A moment later, he turned and strolled across the landing to the small trailer that served as his field office. He went in, removed his tin hat and hung it on a hook, then moved to the cabinet along the far wall and poured himself a mug of scalded coffee from a discolored pot on a propane burner. The brew was rank. He made a twisted face as he set the cup down on his desk.

For a moment he stood deep in thought, rubbing his temples. Opening a desk drawer, he pulled out the contract for the right-of-way project and plopped it down on the desktop which was piled high with unfinished paperwork; unfinished because Jacques dreaded this end of the business almost as much as he hated government red tape, to which there seemed no end.

He settled his large frame into the swivel armchair, snapped on the battery-powered desk lamp, opened a second drawer, fumbled under some papers until he found the Tums, removed the bottle cap and popped three pink pastel pills into his mouth. He washed them down with a bitter slug of coffee, tossed the plastic bottle back in the drawer, and produced his reading glasses. Flicking them open he slipped them over his ears. With the glasses riding low on his nose he slumped back and began to study the contract for the umpteenth time, poring over the fine print, searching for a loophole.

EIGHT

The showdown with the Forest Service happened the following Monday, bright and early. Mickey Newsome and two associates were on time for their 8:00 A.M. appointment at the Sulfur Springs District office. A smiling secretary greeted them as they entered the front lobby and promptly led them into a conference room. She assured them that the ranger would be right with them and graciously offered coffee, but there were no takers.

A few minutes later the district ranger bustled into the room followed by three of his staffers in single file. The ranger immediately turned, as if he had forgotten something, and whispered to a woman who hurried out. After handshakes all around the staffers took their seats along the far side of the table. They were dressed in freshly pressed green-on-green Forest Service uniforms.

For a moment each side silently faced the other.

Newsome had come with low expectations. The chances were slim to none that a spur-of-the-moment meeting such as this would accomplish anything of substance. Ranger Bennett had probably agreed to it only as a courtesy. After all, his planners had completed the scoping process for the Bowen Gulch timber sale months ago. The deadline for public appeals had come and gone; and, regrettably, the environmental groups had missed the boat. After a public offering the sale had been duly awarded to Western-Pacific. The Forest Service was in possession of a signed contract. The timber sale was a done deal. End of story.

Despite this, Newsome was not discouraged because he was not the sort to be intimidated by long odds. He had come expecting a fight.

The ranger fiddled with his tie and said "Alright, we might as well get started. Allow me to introduce myself. I am Doug Bennett, district ranger hereabouts. Sitting on my left is Bill Noonan, my very capable chief timber sale officer." The man nodded. "And to my right is Greg Hansen, team leader for this project ... and, uh, just to be sure we are all on the same page, it's my understanding you are here today to discuss the Bowen Gulch sale. Is that correct?"

Newsome nodded and heard a snicker from across the table. With his long hair and beard Newsome looked more like a hippie than a university professor. He let it pass, however. "Ranger, I want to thank you for agreeing to meet with us. I am Mickey Newsome, chair of the Rocky Mountain Chapter of the Sierra Club. Sitting at my left is Enda Kiley Mills, and to my right is my old friend Skip Martin."

Newsome was by now well aware that he was dealing with three hard-core timber beasts. He knew the type. The snicker and the smug faces told him there was no point trying to engage these men in a serious discussion about ecological principles. It would be a futile exercise; so much wasted breath. During his many years of tracking the Forest Service Newsome had yet to meet a district ranger or a timber sale planner who had the proper respect for ancient forests. Sure, many of them talked the talk, but they invariably embarrassed themselves when it got down to cases. The meeting was moot, anyway. The Forest Service was holding all of the cards. The men sitting across from him knew this, and two of them already had that bored look. One was yawning. Another was absent-mindedly clicking his ballpoint pen. But Newsome had a card up his sleeve. He had watched them closely as he introduced Ms. Mills and was certain they had not recognized her. He nodded to the dignified-looking matriarch sitting beside him.

Enda acknowledged his nod. Her cue. She had silver hair and looked to be about seventy. "I also would like to thank ranger Bennett for meeting with us, today," she said.

"It's a pleasure meeting you. To be quite frank, I am here for one reason, to tell you that my father Enos Mills, who as I'm sure you know was the founder of Rocky Mountain National Park, had always intended that the incredible spruce forest in Bowen Gulch should be included within the park boundary."

Upon learning who this woman was, the surprised men across the table sat up straight and for a delicious moment squirmed in their seats.

She continued, "This was my father's wish. Please believe me when I tell you that the political decision to leave Bowen Gulch out of the park was one of the biggest disappointments of his life. So, I'm here to ask on his and my behalf that you withdraw this ill-considered sale. It should never have been proposed in the first place."

The ranger cleared his throat. "It's an honor to meet you, Ms. Mills. Your father was a remarkable man. Truly inspirational. He is one of my own personal heroes and I'm sure my colleagues here with me feel the same way. However, I must also tell you that I do not have the authority to make policy. My job as district ranger is simply to follow the law as written." He held up a voluminous document and waved it in the air. "I'm afraid you have come too late to the table. The appeal process has already been concluded for this project. It's done. We have a signed contract and it's my job as district ranger to implement it. I'm sorry, but that's how it is."

It was the old familiar dodge, Newsome observed. The ranger was hiding behind his uniform. "So, that's the contract for the Bowen Gulch sale?" he said, motioning with his hand.

"Yes."

"I understand Western-Pacific was the lone bidder and got it for a song. Is this correct?"

"No comment," said Bennett.

"Ten million board feet of high value straight-grain old growth timber for pennies on the dollar, a pittance. So, tell me, ranger Bennett, when did the Forest Service go into the business of subsidizing preferred timber companies like Western-Pacific, at taxpayer expense?" It was obviously a rhetorical question. He continued, "Man, how can you even pretend to hold your head up? Shame on you."

"Now listen here..." the ranger started but he was interrupted by Skip Martin, the third environmentalist who had been listening quietly up to this point. "Do you mind if I examine the contract?" Skip said. "It is a public document, is it not?" Martin was an imposing individual, well over six feet tall, with enormous arms and hands.

"Yes, it is," the ranger said, irritation in his voice. "By law anyone may see it." Reluctantly he passed the contract across the table.

The file was at least a half-inch thick. Martin flipped through it briefly, then, held it up with his big hands and ripped it in two with no apparent effort. He tossed the shredded document onto the table before the ranger. "So much for your signed contract," he said with a wry grin. "Back to the drawing boards, gentlemen."

With that the meeting broke up in noisy disarray.

Later, Newsome and his comrades re-grouped at a local tavern. Pinecone Peters had joined them. "I only wish I could have been there to see the look on Dougie's face when you ripped up the contract," he said. "If I sound like I'm gloating it's because...well, I am." He laughed. "It's personal, y'see. I could tell you stories about my old boss, believe me. Too bad none of you had a camera."

"It was a hoot," agreed Skip. "But the only thing that matters now is how do we stop this damned timber sale."

"Yes, how?" said Enda. "They have awarded the contract to W-P. What legal options do we have left?"

"I'll have to check with our attorneys," said Newhouse. "However, my guess is that, at this point, we have none. No remaining legal options."

"So, then, it's over?"

"I didn't say that. Hell no; I believe we can turn it around. We can still win this fight. We have to. Not through a legal challenge, but in the court of public opinion. Swing the public to our side and we'll win. My sense is that the citizens of Colorado would be outraged if they knew the facts, what the Forest Service is planning to do."

"I agree," said Pinecone. "The way to win is to tell the story of Bowen Gulch. That's what I've already been doing, or trying to."

"But how do we reach the public in time?"

"Obviously, we have our work cut out for us," said Newsome. "Folks, I can't speak for the rest of you, but in all of my life I've never backed away from a fight. Provided I felt the issue was important enough..."

Skip chimed in, "Right. So, how do we feel about Bowen Gulch? Is this place *worth* fighting for?"

"Absolutely," said Pinecone.

"Yes, definitely," said Enda.

"Then, I'm all in," said Skip.

"Alright," said Newsome. "It's decided, then. We agree. The campaign to save Bowen Gulch starts here and now."

"I was up there the other day," said Pinecone. "There's still two feet of snow in the woods. It's going to be too wet to log at that elevation for another month. The woods probably won't open up until August."

"That gives us at least a month."

"Not much time."

"Delay is our friend."

"We'll be ready."

"We'll have to."

NINE

The wind-storm that shut down St. Clair's right-of-way operation on Cameron Pass the previous October was – well, there are no words adequate to describe it this side of hell.

The project amounted to a huge 350-acre clear-cut preliminary to groundbreaking for the new catchment reservoir at the headwaters of the Poudre River to back up snowmelt.

"It's a big state job," Jacques had told his crew. "Stage one, that's us, was suppos' to be done last month. But with the breakdowns we've had, an' one thing and another, we're running behind schedule. I figure 'bout two months."

The boss laid it out. In order to catch up he wanted them to work straight through, seven days a week until completion, "or until 'old man' winter drives us out of the woods. Whichever comes first. You never know. With a bit of luck we might git'er done. Some years we work into December."

He had no trouble bringing them around. The boss spoke their language. His men were looking ahead to the winter layoff, and the promise of a big fat paycheck just before the holidays, including a $200 bonus for each man was too good to pass up.

"It'll make things go easier with my wife," said Kermit.

"You mean your ex."

"Hah! With the child support he's paying he'll be lucky to see a dime."

The valley was heavily forested, but access was good because the site paralleled the adjoining Cameron Pass road. Jacques instructed them to start at the northeast end, the narrowest point and the site of the future earthen dam. No sweat. In short order they cleared the area of trees, then began working methodically up valley.

At that point Jacques brought in his cat and began to construct a landing. Meanwhile, Francis went to work with one of the skidders.

For many weeks the mountain weather had been favorable. The early October nights were cold, with hard frosts every morning; but in the afternoons the temperature climbed into the fifties, some days into the sixties. In short, the brisk mornings and balmy afternoons were ideal for autumn logging.

A chainsaw is a powerful instrument in the hands of man who knows how to use it. A dozen experienced cutters can make short work of a stand of timber under contract of the saw. Within a week the crew had ripped the heart out of the stand; whereupon, they fanned out to work the perimeter, flagged by yellow ribbons. Each day, the men descended on their timber allotments like a horde of hungry defoliators.

By week's end, the forested valley resembled a moth-eaten tapestry. The clearcut sprawled over at least 140 acres and was expanding by the hour. It was the largest stump field Tom had ever seen.

Jacques was pleased with the progress.

Then one morning they awakened to a red dawn, the first hint of a possible change in the weather. By noon the sky had clouded over, and the following day they never saw the sun. The next morning, Tom was making ready for work in the cold predawn when Mike Garity appeared with his saw slung over one shoulder. He needed a lift.

"Sure," said Tom. "Hop in."

"About froze my ass off, last night," Mike said as he stashed his gear in the back. He paused to consider the overcast. "Looks kind of grim," he said as he slid into the cab.

"Yep, something's brewing."

"Boss's gonna be hell on us if we don' get 'er done before the snow flies. Lord knows I need the cash."

Tom pointed to the far ridge, a swathe of fall colors that ranged from brilliant yellow to orange to red. "Check it out. The aspens are peaking."

"Nice. You hear a forecast?"

"Nu-huh."

Garity clicked on the radio and worked the dial. There was plenty of country music but no weather until the top of the hour.

"It's twenty til."

"Screw it."

"Where do you want out?"

"The main landing."

"Isn't that your blue Chevy?"

"Yeah. It wouldn't start, last night. Got a plugged fuel line I reckon." Mike removed his gear from the back. The logger's face loomed large in the rider's window. "Thanks, kid. Appreciate the lift." The man banged on the door, two times. "Hey, don't work too hard."

"So long."

Tom doubled back to his allotment. The work was routine, until about mid-morning when a faint breeze rustled through the forest. It was no more than a whisper in the treetops. Tom was too busy to notice. But the zephyr-like breeze soon gained in strength and, quite suddenly, boughs were rustling all around him. The entire forest seemed in motion, trees swinging and swaying, this way and that.

The breeze was now a hard wind and still rising.

Shortly after dawn, the stationary high that had prevailed for many weeks over the Rocky Mountain Front had begun to move eastward when a trough of low pressure, which had been developing 100 miles east of the Rockies, reached a critical threshold.

At that point an ocean-sized mass of air suddenly began to shift as one across thousands of square miles, gathering above the state's northern cordillera, the high granite peaks and red stone mountains, above the highest summits and deep snow-fields and along the crested divides. The great tide of air came howling down through the mountain passes and rugged defiles, pressing eastward, rushing to fill the zone of low pressure out on the high plains. They say Nature abhors a vacuum. Believe it.

The wind dropped off the jagged backbone of the Gore and Park Ranges and came funneling through the mountain passes, Cameron, Willow Creek, Milner, Berthoud, Loveland, each a natural wind tunnel, whistling down the snaking valleys of the Front Range at forty nautical miles an hour miles an hour and still picking up speed.

The fast-developing windstorm soon topped sixty miles an hour, gale force, scouring the treetops like a million freight trains.

As the tempest rose, so did the roar, until the shriek of the wind became a screaming crescendo that drowned out the puny wail of chainsaws.

One by one, Jacques' loggers shut off their saws and stood watching the freakish wind that had enveloped the forest around them. What else could a man do but watch? It was too dangerous to work under such conditions.

Along with the gale now rose a blanket of air-born dust and debris. Weirdly, it appeared to hang suspended above the valley like a yellow apparition, an apron of dust that muddied the skies and darkened the already dim sun as it spiraled in broad eddies above the treetops.

Here and there shallow-rooted lodgepole pines began to go down with a sharp wrenching and snapping of roots.

As the wind gusted to seventy miles an hour, earth and sky became a maddening blur of motion, a million points of distraction. The devilish wind was its own species of conundrum.

Cyclonic dust wormed its way into Tom's eyes, gouging his ears and face like stinging nettles, almost blinding him. Around him the forest had become a chaos of noise and movement. Tree crowns and branches whipsawed crazily, every which way.

The crew had already carved a broad swath of stumpland through the forested valley, amounting to about half of the 350-acre right-of-way. In places, only a narrow strip of timber remained on its feet. Miles of forest "edge" had also been hacked through the remaining timber, all of which now stood unnaturally exposed. Upon the anvil of this remnant forest the deafening gale now fell like a hammer.

As the blow surpassed seventy miles an hour, deep-rooted centuries old mossy-barked firs and behemoth spruces began crashing over. Soon giant trees were going down like dominoes. A big trunk would crash against another tree, *khe-thunk*, triggering a chain reaction – an explosion of popping roots and broken limbs as trees shin-boned one against another. Sometimes three or four tangled trunks would go down together in a collapsing shower of broken branches and flying bark.

On every side, ancient trees heaved and groaned like wide sails straining to the limit before the mast, and beyond.

Tom initially had lowered his head to the storm, gritting his teeth as he attempted to stay focused on the work. But when a big tree went down too close for comfort, he packed it in. It was time to get out of the woods. Hastily he gathered up his gear and made for his pickup. But it was not easy going. The footing was terrible. The skid trails now were non-functional, blocked by a jumble of downed trees jack-strawed every which way. Stems and branches lay piled in heaps; and the wind was still rising.

The loggers were a sight as they scrambled to safety, picking their way through the slash and fallen trees, dodging falling debris as they struggled to hang onto their gear and tin hats. The men's heads kept bobbing up and down as their eyes darted back and forth from the uncertainties underfoot to the dangers above. Their saws were balanced precariously on their shoulders. It was a perilous business.

Jacques waited on the landing. Like the rest, he had one hand firmly on his hard hat. The boss acknowledged each one as they came in.

"Anyone see Mike?" he said, squinting into the storm.

"He's coming, boss. He's right behind me."

"OK."

From the relative safety of the landing the men watched like dumb statues as the forest disintegrated before their eyes. The timber was being torn to pieces. Awe and fear of the unbridled power of Nature was etched on every face.

"Oh my God! Oh my God! Oh my God!" one logger mumbled over and over again, like a broken record.

After awhile they picked up their saws and quietly returned to the trucks. There was nothing to do but retreat to camp and wait it out. Wherever the forest canopy remained intact the forest itself offered a measure of protection.

But their trials were not over, not yet. On the way to camp they were compelled to clear fallen debris from the road using their saws and, on several occasions, a winch.

The Preacher came within a whisker of not arriving when his truck sustained a direct hit from a falling lodgepole pine. Luckily, the roof on the rider's side took the brunt of the impact. Dipstick was luckier still. At the last moment he had decided to hitch a ride back to camp with Jimmy instead of with the Preacher and so averted a terminal headache.

The near miss turned weirdly comical when they found the Preacher standing in the road waving his arms and talking animatedly to himself, as he was wont to do. "She was in pretty poor shape to begin with," he nervously pointed out, apparently trying to sound optimistic. It was true enough. The old truck was a jalopy. Jimmy and Dipstick cut the lodgepole free, dragged it off in sections and dumped them along the side of the road.

"See if she'll turn over," Charlie said.

"Yeah, give it a crank."

The front windshield had partially popped out and the door on the rider's side was tweaked and wouldn't close. But when the Preacher – he was jittery – finally settled down and climbed inside the motor turned over and sounded just fine.

"Hoo-ray!"

"What do you know?"

"How about that!"

The cab was a problem, however. The crumpled roof made it impossible to sit upright in the cockpit.

"That'll give you wry neck."

The Preacher rummaged around in the back and began pounding on the roof from the inside with a ball peen hammer. When that failed to produce results he lay on the seat with his big clodhoppers in the air and kicked the roof up; not much but enough that he could sit behind the wheel without scraping his head. He then pounded out the rest of the windshield.

"It's gonna be drafty."

The vehicle was a wreck. Any other man would have written it off. But the Preacher showed every intention of pressing it back into service. The situation was beyond absurd and the bystanders could not resist wagging their fingers at him as they offered advice.

"Now fix that flat and you're done," said Sourpuss.

That evening, the crew made hilarious small talk of the Preacher's cockamamy rig. The tomfoolery had a sharp edge of mockery.

The windstorm finally played out during the night.

But no one was laughing the next morning when they discovered Mike Garity's rig still parked near the landing. They found him fifty yards from safety. Mike was sitting on the ground leaning against a fir tree, both legs extended. His eyes were open, set in a perpetual stare, a slight frown on his face. Otherwise he looked quite relaxed, as if he had sat down to rest a spell. A three-foot shard of wood was sticking out of the top of his skull. An ugly line of coagulated blood ran down one side of his face.

"Now, he can rest for eternity," stammered Thurston. "The poor boy never knew what hit him."

"Nope. Probably never felt a thing."

"Damn. He almost made it."

"Wasn't wearin' no hard hat."

"It wouldn't a' made any difference," Jacques said quietly.

"Yeah. Broke his neck sure."

Mike's chainsaw was on the ground beside him, his hand still gripping the handlebar. The grip was so fierce that when the time came to move the body they had the devil of a time separating the chainsaw from the man.

"You know," said Charlie, later, "it's almost as if old Garity wanted to take his Husky with him into … the hereafter."

"Yeah, like he knowed he was going to need it."

"Damn it all."

"Makes you wonder, don't it?"

There was more wind that evening. Later, it turned bitter cold.

Sometime in the night Tom awakened with a case of the shakes, his sleeping bag soaked with sweat. Of the bad dream he remembered nothing apart from the stench of death – and the vacant eyes that gave him no peace.

TEN

Their affair in Florida had been short but intense, six days shacked up in a motel room. Later, Tom often wondered about her, and why she picked him. After all, he had failed to defend her honor.

But for his decision to go south, they would never have crossed paths. He certainly had no plans to winter in Florida, that year. The trip was born of necessity, the mistress of invention.

After the early fall shut down, Tom found himself casting about for work; just another unemployed logger with grit under his nails.

Only there *was* no work. The economy had tanked in September. The country was in the midst of the deepest recession in half a century; in part, the result of a worldwide oil glut.

The rest of St. Clair's crew took the easy out. The loggers kicked back to collect unemployment checks until the spring thaw hopefully brought better days, all of them, that is, except Tom Lacey, who failed to qualify for relief. He had not yet paid enough into the general fund to meet the minimum state requirement. He came up just short.

So, he joined the ranks of other out-of-work men and queued up for day labor. He also scanned the job boards, hands in his pockets. For two weeks he battled boredom waiting for something to turn up.

Finally, something did. From a posted handbill he learned about a treeplanting company based in Fayetteville, Arkansas. The outfit was recruiting for large planting contracts in Florida and Georgia. The poster read: "No experience necessary." Tom felt he might as well try for it.

He put a call through to the home office and was hired on the spot. Soon, he would be back at work. Start-up was the twentieth of November.

The only remaining issue, the matter of shelter, was solved when he chased down a used camper shell. He picked it up for a song, mounted it on the back of his pickup, rigged up a two-burner cook-stove, installed an ice cooler, and *presto!* His truck had become a house-on-wheels.

Tom thanked his stars that he was footloose, without family ties and obligations, free to follow the work, wherever. He put his saw and gear into cold storage at Red Callahan's – Red lived in the Fort; and the night before he left, he gathered with some of the men at the Town Pump for a send-off. They thought the treeplanting gambit was hilarious.

"Tom the treeplanter," croaked Shorty. "Hehehe.."

"Yeah, Johnny fuckin' Appleseed!"

"He's going to even up accounts," said Dipstick. "An' pay off his karmic debts."

"Bullshit," Tom said. "It has nothing to do with karma. I have to eat too, same as you bums."

They toasted him a last round.

"Here's to the scholar."

"Nobody gets out alive," said Charlie.

"Don't worry. Tom'll be back. He loves dropping them big ones..."

"Oh yeah he does," said Red.

The heads nodded. They knew the kid would return in the spring. He was hooked; once a logger, always a logger.

He pulled out of Fort Collins on the tenth of November headed east, start of a week-long road trip. He spent three difficult days in Amarillo with his last surviving uncle, T.R., who was failing after a series of debilitating strokes. After paying his respects he said good-bye, they both knew for the last time, and hop-skipped to Springfield, Missouri for a reunion with former college friends. Two days later he spent a night in Montgomery with his crazy aunt Catherine who had hounded three husbands into early graves. The last night he slept on a lonely beach outside Destin, Florida. The rest was a jaunt. He made the work rendezvous at a designated state park east of Gainesville, Florida, with time to spare. The temperature in sunny Florida was a balmy sixty-five degrees, shirt sleeve weather; the first of many surprises to come.

ELEVEN

Tallie was in the crowd that first morning, though Tom failed to notice her, when a lanky man with a beard stepped up and introduced himself.

"Good morning, good morning. My name is Ed Conyers and I hail from Arkansas. From up around Fayetteville. If you are familiar with that country you know that up there we're *all* good old boys." He paused to twitters of laughter. It was good form, Conyers knew, to season the straight-talk with humor, especially on the first day. No point in scaring them off.

He continued, "On behalf of Reforestation Incorporated I want to welcome y'all to the wonderful world of tree-planting. It's going to be my job to serve as your crew boss. For the next few months we'll spend most of our time together. The season will run through next March. I'll tell you what, it's going to be an adventure and a' expect by the end of this week we'll be one big happy family around here. Whose golden retriever?"

"Mine."

"And who are you, suh?"

"Bill Nelson."

"Pleased to meet you. Nice lookin' dog."

"Thanks. His name is Amazon."

"Where you from, Bill?"

"Madison, Wisconsin."

"Good to have you aboard." Conyers continued, "By my count there ought to be nineteen of us, twenty countin' Amazon." A lone laugh.

Conyers did a roll call. As he went through the names each one raised an arm or said "Here!" or "Yo." The crew included three women.

As he looked them over Conyers was already having reservations. Several men had returned from the previous season. Thank heaven for that. Experienced planters made his job easier. As for the rest, they would not know which end of a seedling was up. He would have to start from square one. It was obvious that most of the new recruits were on the rebound from the bad economy. Treeplanting required no previous experience and so, was a last resort for men down on their luck.

Conyers frowned as he looked them over. They ranged from long haired hippies and back-to-the-land types from the coast of Maine to more conservative men from the deep south, including several rough-looking drifters. Ne'er-do-wells. He knew the sort. He had seen their like before. Too many times. Such men made poor treeplanters, and often spelled trouble.

"OK. Looks like we're present and accounted for," Conyers said in his slow southern drawl. At a nod, one of the veterans started passing out the standard issue equipment. In addition to the hoedad the gear included a pair of tree bags to hold the seedlings. The double bags were part of a belt-and-shoulder harness designed to hang at hip level.

"The bags and hoedad will cost each of you seventy-five bucks," Conyers told them. "That will be docked from your first paycheck. Any questions?"

Seeing there were none he proceeded with his harangue. "OK. Let's see a show of hands. How many of you have used a hoedad, before?" Only three hands went up. "Three. That's what I thought. For the rest of you, I know it's kind of strange looking. But as you are about to learn, a hoedad is an extremely well-designed tool, very functional."

It was the industry standard planting tool, and resembled a garden hoe, with a much longer blade and a tapered handle. The thing was odd-looking indeed, until a man learned how to handle one, which took all of a few hours.

Suddenly Conyers was brandishing a hoedad before the group, swinging it around in a full range of motion. He did it with practiced ease. "These are nicely balanced tools," he said as he passed the hoedad back and forth from hand to hand.

There was something sensual about the way he ran his hand up and down the long wooden shaft. "I love the feel of these oak handles," he said, "and I'm sure you will too. Before the week is out, your hoedad is going to feel like part of your own body. Now pay close attention." There was no need to say it. By now, all eyes were glued on the boss.

Conyers swung the tool overhead as before and, in a smooth fluid motion that never broke form, brought the blade down hard and fast and buried it in the sandy ground up to the neck. Bending down, he worked the hole open and made it look easy. Next, he reached into one of his harness bags and pulled out a small seedling. He held it up high for everyone to see. The roots dangled below the small stem, dripping wet. Conyers stroked the mass of wet roots with affection, as if he were fondling his own beard.

"OK, we'll run through it. This is a Loblolly pine seedling, and this here is the root. First lesson. Tree roots grow down, not up. Understand?" Nineteen heads nodded. Conyers pointed to the lowest tip. "This little fella, it's called the leader. This is the most important part because it's the growing point. This little guy needs to point down. I mean straaaaight down. You all hear me?" he almost shouted. "What did I just say?"

Someone volunteered, "You said, point her down."

"Right," Conyers said. "Straight down. I want him headed for China. Otherwise, the seedling just won't make it. Now, watch." Conyers twisted the leader up into a "J." "See this? This is a no-no. This is called a 'J' root and this we can't have. This, I promise, will get you fired. Because, like I done told you, and, folks, I'll be repeating it until you are sick to death of hearing it, tree roots grow down. THEY DON'T GROW UP. Get it?" Nineteen heads nodded in unison.

"Now, watch this." Leaning over, he placed the seedling in the hole, just so, with the tip down. Then, he closed the hole with two quick swipes of the blade. A final stomp with his boot packed the soil firmly around the tree. Voila! He was done. As Conyers rose he gave the seedling a playful stroke.

"I want you to tamp *every* tree with your boot. Understand? Tight as a virgin. That way there's no air pockets."

Now he tempered his voice. "I need consistency. The key is to make your hole deep enough, and to place the seedling correctly. Do it the right way each and *every* time and there won't be any problems. It's not hard, believe me. It's easy. If you are sincere and I catch you "J" rooting trees, I will work with you on your technique until you get it right. I'm a patient man. Heck, I'm easy as pie to get along with. You do right by me and I'll return the favor and do right by you. But I have to warn you. If any of you have any ideas about pulling a fast one, you better think again. Do not go there. Because I know all of the tricks. Oh, you might fool me once or twice, but in the end I will bust you." He paused to let the stern words sink in, then continued in a milder voice. "We had a fellow last year, a wise guy from Texas, I don't know, maybe that was his problem, who dumped three bags of seedlings out in the woods and took credit for planting them. He snookered me for a week, but when I nailed his ass he was out'a here. The wise guy ended up paying for three thousand seedlings, which I docked from his last paycheck. So my advice to all of you is, get in the habit, from the start, of doing it right. Quality is everything in this business, and I mean everything. We can't have sloppy work. I won't tolerate it. My number one responsibility is quality control and, starting tomorrow, the company rule will go into effect. Three strikes and you're out. Three strikes won't apply today. No, today is for learning. Today, you'll practice the moves and get them down. Heck, by tonight I'm sure most of you will be expert treeplanters. Do I make myself clear?" The heads nodded assent.

"OK," Conyers said, clapping his hands. "Thank y'all for listening. Let's load up. Time to plant trees!"

So it began.

On that first rough day, everyone suffered from sore flesh. There were stiff muscles all around.

About mid-morning, Conyers had to let one man go. Tom never did learn his name. The guy was a fatty, and a plodder, with a peculiar shuffling gait. He moved with a waddling roll of the hips, pants riding low. It was a miracle they stayed up at all. One of the veterans whispered, "If that guy moved any slower he'd be a statue..."

Conyers took the poor fellow aside to deliver the bad news. They never saw "the waddler" again.

But he was the exception. By the end of that first day the rest had more or less mastered the use of the hoedad and the art of planting. The skill came through doing. How do you get to Carnegie Hall? Practice, practice, practice.

Within days the crew was toughening to the work.

The routine never varied. Each morning, the boss rose early and transferred a day's supply of seedlings from a humming electrical refrigeration unit parked in the woods into the back of his pickup. The seedlings were packed in large brown paper bags and had to be kept cool round-the-clock to keep them viable. Temperature and moisture were critical. Each morning, shortly after dawn, Conyers would be waiting for them when they showed up for work. The planters would rip open the paper bags and stuff as many baby trees as they could carry into their pouches, at least a thousand at a time. After loading up they would move out.

It was "piece work," three cents a tree.

The planters would form up in a long straight line that became a diagonal formation moving across the landscape. The standard distance was twelve feet between seedlings.

The wedge formation was simple but effective, enabling the planters to cover the maximum amount of ground in the shortest time.

Soon they gravitated to their places. The quick and the determined moved forward to the head of the column. The slow, the methodical, and the weak slipped to the rear.

Two of the women quit before the first week was out. One was a husky earth mother type named Linda Swaggart. She was an easygoing extrovert and not in the least bit bashful about why she quit. She hated planting and told them so to their faces. She even mocked them. "I think you're all nuts," she said with a laugh as she tossed her braided black hair back over one shoulder. Linda was up front about everything.

Tallie, the other woman, never gave any reasons, but did not have to. The reason was plain enough. She was slight of frame, just a slip of a thing. Tallie could not have weighed more than 110 pounds and did not have sufficient body-mass to shoulder forty-pound tree bags. She was a beauty though, with fine-boned features, and a lilting, childlike voice. Because of the way she was put together her clothes seemed too large for her and hung loose on her small frame. She had her own way of moving, which some would call ungainly, even awkward; but she also showed flashes of a natural grace. The overall effect was stunning. Though she was generally quiet, on occasion she could be loquacious, a curious combination. She always wore sunglasses, even at night.

To generate income and keep busy during the long days, Tallie and Linda organized a community kitchen, pooling their money and cooking skills, figuring to cash in on a commodity in short supply: home-cooked food. The field kitchen was an immediate hit because the food was out of this world. And the price was right, six bucks for all you could eat.

As the roughshod planters lined up, one evening, for their first community meal, the talk in camp was a droll medley of voices.

"Tell Daugherty to cut it out!"

"Tell him yourself."

"I did. But he don't listen. Maybe if you tell him."

"Tell him what?"

"Oh, Daugherty's picking his nose again, and it drives Ferman out of his gourd."

"So what else is new?"

"Now you mention it, he gets it up there don't he?"

"Yes he does, with them long bony fingers."

"If he don't watch it he'll puncture his brain."

"Brain? What brain?"

"It ain't the picking that bothers me," said Ferman. "It's what he does with it, *after*."

"Jesus! Will you look at that!"

"Oh Gross!"

"Daugherty, you're a disgrace to your momma."

"He obviously don't care about himself, or nuthin."

"Booger Daugherty." Chuckles all around. Tom laughed with the rest as he slapped at a mosquito on his neck. Others stepped up now and joined the food line. Each had a plate, a cup, and silverware.

"Smells good. What *is* it?" one man said as he handed Linda six bills.

"Yeah, what's for supper?" said a face peeking over John's shoulder.

Linda was beaming. "Onion soup," she said. "Homemade corn bread with butter. Fried chicken. Mashed p'taters and gravy. Half a chicken for each one of you. A big pot of peas. Oh and uh, Tallie made a peach cobbler." The planters smacked their lips.

"All right!"

Tallie began serving up the chow. When Linda finished collecting the money, she helped too. They piled them high. Each man filled his cup with soup or bug juice, then found a spot in the grassy clearing. They ate in ravenous silence. But when the edge was off their hunger, the small talk resumed between mouthfuls of chicken, spuds, and peas.

"Sometime early next week we'll finish up the parcel," said Conyers.

"Where then, chief?"

"We got two more sites, hereabouts. Then we go south."

"South where?"

"Daytona."

"Alright by me. I'm a beachcomber."

"Mmmm that cobbler's gooood. Is there more, Linda?"

"You'll have to wait till everyone's had firsts."

"We won't be on the coast. We'll be working 'bout fifteen miles inland. Couple of Georgia Pacific contracts."

"Hot damn. We're gonna be rich!"

"Rich. Hell."

"At three cents a tree, boy, you ain't never gonna be rich."

"No, we gonna be. I know it. I dreamed about it, last night."

"Right, Watters. Dream on."

"I think I'll get me some of that cobbler. Is there any cream?

"There should be. Look in the cooler."

"So what did you dream, Jamie?"

"Man, I dreamed we was planting money trees. As in M-O-N-E-Y! It was unreal."

"Money trees. Hah!"

"Unreal. He got that part right."

"Sounds high on the hog to me."

"Can you imagine? Pickin' hunnerd dollar bills..."

"The water's hot," Tallie said in her musical voice. After serving up the food the women had set two large pots of water on the fire. One was soapy and the other clear for rinse, so they could wash their dishes and also purge some of the day's dust and grit.

Heads turned when Tallie moved away from the fire, her long calico dress gathered up in both hands.

Next evening,Tallie and Linda caused a stir when they were seen frolicking together, laughing, embracing and kissing on the lips. There was nothing casual about the display of affection. When Conyers heard about it his reaction was "Oh shit." The crew included some conservative southerners and he could not help but worry about the possibility of trouble. Once before, he had seen a crew split down the middle over something as trivial as an off-color remark. Fortunately, nothing much happened, aside from a few cold stares.

Next morning early, the boss caught two of the ne'er-do-wells "J"-rooting pine trees. The two had already had their first warning. This time, Conyers was really pissed off and whistled everyone in. The planters trudged over and gathered around. The offenders endured a verbal spanking with their heads down and their eyes on the ground.

"If I catch you boys 'J'-rooting trees again you'll be gone. Understand? Out of here. Strike three and you're out."

"Sorry, boss."

"Sorry don't count."

"We'll try to do better."

"I need results. Look, you guys are moving way too fast. Slow down and do it right. Focus on quality, first. Then, pick up your speed. It will come."

"OK, boss."

After that, the two men settled down and, for a time, there were no more problems with 'J'-rooting.

The next Friday, after ten straight days working from dawn to near dusk, "can to can't" in treeplanting parlance, the crew dispersed to Gainesville for the weekend and some needed R&R. It was a fun-loving college town, with lots of night life and good restaurants.

One group, including Tom, designated Bill Nelson as 'point man' for the expedition, because he was the most presentable. Bill still boasted a somewhat clean shirt and had managed a recent shave. While the others held back, he signed into a Ramada Inn as a party of one, politely requesting "...a double suite in the back, away from the street, so I can stretch out and relax. I need some sack time..."

Within minutes of check-in six grungy planters and a golden retriever descended on the room with six-packs in hand, tracking in sand, grit, mud, mange, fleas, and swamp reek; a motel maid's nightmare.

Sprucing up assumed top priority. The men queued up for hot showers and shaved at the mirror, while the rest guzzled cold beer and watched cable TV, a process of hygienic restoration that soon expended every article of clean linen in the place. Promptly they proceeded to the next order of business, dinner at "Gators," a local eatery; after which, the men returned to the room to laze away the stag weekend watching round-the-clock cable movies and pro football, while scarfing endless junk food and swilling beer by the case.

By 11 p.m. Saturday night, the topic of conversation had turned to "women." Things were beyond mellow. Bill Nelson lay flat on his back on the floor on the verge of an alcoholic stupor. Someone held a can of Bud over his head and emptied the suds onto Bill's face. Some of the beer made it into his mouth. Most did not, but no matter, it was all the same to Bill. "That's when she smiled at me," he moaned, "with them beeuutiful brown eyes." Amazon barked playfully and began lapping his face. "Hey? Whaaa? Iye aaawww..."

"Nelson's got a thing for Tallie."

"So *that's* what he's been raving about like a loonie."

"Good luck with that one."

"She likes to have women suck her clit."

"That boyfriend of hers. Alan. What's with *him*?"

"He ain't no boyfriend, that's her cousin."

"Well, what about Linda's boyfriend? Ned. What's he think about it?"

"Oh Ned don't care. He's laid back."

Bill's mumblings were becoming incoherent as the tail-wagger continued its ardent ministrations.

"Nelson says Tallie smiled at him. One time, that's all she wrote. Now look at him, a good man turned to mush."

"She *is* a cutie. I like the way she talks."

"What a waste of good pussy."

"Well, I wouldn't kick her out of bed, that's for sure."

Tom Lacey was probably the only man in camp with no strong opinion about Tallie one way or the other.

Come Monday they were back at it, hung over but refreshed nonetheless after the wild weekend. The morning turned into one of those good and glorious days when everything is right with the world. The planting seemed effortless. The tree bags never became burdensome. There was a welcome breeze – the mosquitoes never showed up. There were no eastern diamondbacks coiled in the brush and no pestilential thistles ripping their flesh. The crew was in "gravy" all day, flat sandy ground. There was joking, free and easy laughter and a feeling of camaraderie, especially among the group that had coalesced around Bill Nelson.

Tom, who loved dropping trees, had by now taken to planting them with near equal enthusiasm; perhaps because it was hands-on work. The silent, earthy rhythms were so very different from the noisy mayhem of logging; yet, Tom found hefting a hoedad to be immensely pleasurable. The best part was the pure elation born of mindless routine, when monotonous repetition set him free on his feet to think and to imagine.

He had one such experience later that very afternoon, near dusk. Down to the last few seedlings at the soggy bottom of his tree bags, the sun riding low, a ball of molten fire in the western sky, the air already feeling cool against his skin. As he paused and watched the sun slowly flatten down into pure ochre, then slip behind a pastel thunderhead piling up over the Gulf, something stirred within him, an inner fire. Everything shifted then and suddenly he was in a different time frame. The work had become a passion play without beginning or end. He was dancing with every swing of the blade, light of step, in tune and rhythm with everything, totally in the here-and-now of Nature and rolling up the past. Each thrust into the ancient ground was an act of co-creation and each boot-tamped seedling a rite of renewal. More than this, his hoedad had become a vajra blade, the razor edge of discrimination. With every swing he was shaping the world to come, carving out an arc of future possibilities limited only by the reach and integrity of his own thoughts.

It was a rare moment of pure elation, and one that renewed his confidence in the course he had begun to chart – the path of inquiry.

The evening concluded with a torrential downpour accompanied by hell's own fury, berserk lightning unleashed like wild artillery. Again and again the western horizon uncoiled with incendiary flashes, barrage after barrage, in rapid series with delayed thunder rolling and booming in the purple distance.

This strange delight of winter lightning was new in his experience.

He read only one book that winter, *The Critique of Pure Reason* by Immanuel Kant, and he spent many an evening with it, alone, by the light of a lantern, propped up against a pillow in the back of his pickup, serenaded by the frogs and the crickets in the pine brakes around camp; and, despite terminological difficulties of every kind – Kant is by no means easy – he felt that he made steady progress.

Tom was always the last to douse his light. As the days passed Bill Nelson discovered his habit of reading late, and began to kid him about Kant, always in a good-natured way.

One evening Tom was reading by the light of his lantern, as usual, accompanied by the roar of a billion croaking frogs, when someone lightly tapped on his camper window.

"Who's out there?" He looked closer. "Tallie? Is that you?"

"Yes, it's me. Hi, Tom. Can you help me? I ... umm ... think I need a light." She explained that she'd ventured out into the woods to pee, but her flashlight batteries had gone dead. Conyers had emphatically and repeatedly warned them about the danger of walking around after nightfall in cottonmouth country.

"Umm, sure," Tom said. "No worries." He opened the back hatch and climbed out with his Coleman. He turned it all the way up to expand the circle of light and escorted her back across the clearing to her trailer. They had not gone ten yards before they encountered a large cottonmouth coiled in the grass. The venomous white mouth drew back with a faint hiss as the fangs leered up at them. The serpent seemed primordial – something dredged up from the depths. That was when he noticed she had no shoes.

"My God. You're barefoot!"

She smiled.

"It's dangerous."

"But I love the grass between my toes."

He let this bit of blithe insanity pass without comment and they detoured around the snake. When they reached her trailer he lifted the lantern up to get a better look at her. It was the first time he had seen her up close absent her sunglasses. She had freckles on her nose, yes, and suddenly he knew why some of the men called her "peek-a-boo." Her beauty flashed at you, now you see it, now you don't. He stared at her. He couldn't help it. Something in her dark eyes startled him. Before he could say "Good night," she wrapped her arms around his neck, pulled him down and gave him the sweetest kiss. She smelled really good. For a moment he felt her looking into his soul.

"Thank you, Tom. You're a life saver."

"N-night," he stammered.

TWELVE

The next day, about mid-afternoon, the crew completed the first big planting contract of the season. When they returned to camp they were surprised to find a succulent meal of barbecued spare ribs waiting for them courtesy of Ed Conyers who had arranged for Linda and Tallie to baste the ribs southern-style over a slow fire. According to Conyers, the sauce and charcoal fire was the secret to "finger-licking-good" ribs. He claimed the sauce was an old family recipe. The women had also prepared a large salad, two different vegetables, baked potatoes and endless chips. Conyers had brought in a keg of ice-cold Bud, which immediately drew a thirsty crowd around the tailgate of his pickup.

When the crew had done making short work of the ribs, Bill Nelson produced a guitar and one of the southern boys appeared with a fiddle, which he proceeded to tune up; and suddenly there was music by the fire and lively dancing. Linda and the other woman, whose name was Carol, danced with their boyfriends for awhile; then, Linda grabbed Tallie by the arm and swung her around. The two danced cheek to cheek, making it up as they went along. After awhile, though, they split up and danced with the men, even the southerners, who welcomed the opportunity to cavort with beautiful women, even if they were lesbians.

Before the evening was over, Tallie and Linda had danced with them all, everyone that is, except Tom, who had been in a sorry state since that lone kiss the previous evening. Unable to sleep, he had laid awake for hours with his hands beneath his head contemplating all manner of things, a slip-stream of thoughts and emotions that did not play out until shortly before dawn when his eyelids finally closed. All day at work, he stumbled along in a stupor, hardly conscious of what was happening around him. Fortunately, by now, the moves were second nature. He planted on autopilot most of the day and did well enough except that, one time, he missed the turnaround at the edge of the parcel, marked by bright flagging, and continued in a straight line onto the adjoining property. Fortunately, one of the planters shouted him back. No harm done.

Tom had been with women before but labored under the illusion that he knew something about the opposite sex. Now that he had been rudely awakened to the reality, he was in a muddle. Once, when he saw Tallie smiling at him he thought "Oh my God if she comes over I'll probably lose it completely." He had no idea what he would say to her. He worried that if he opened his mouth and said one word, he would start babbling like an idiot. He was tongue-tied and star-struck both.

Yet, Tom had never been so happy.

When the dancing played out, they sang songs around the fire for awhile, then, told jokes and funny stories. The feeling was mellow, with laughter and kidding around. The fault line between the new age folks and the southern contingent had softened. Linda and Carol snuggled with Ned.

One of the southerners said, "You sure don't *look* like..."

"*What?* Like what?"

There was a pause. "Never mind." Someone guffawed.

The evening ended after midnight when everyone said their good nights and drifted off to tents and trailers. Tallie and Linda left hand in hand.

Conyers was well satisfied.

Little did he know that his "big happy family" was about to crash and burn.

Next day, Conyers moved the crew seventy-five miles south to the swampy outback west of Daytona where cypress forests were being "reclaimed" and converted to pine plantations. The sites were so soggy that it was necessary to first prep them at considerable trouble and expense. Standing water had been drained and raised beds of soil mechanically dredged up to prepare the ground for tree-planting. In this way the owners evidently figured to bring "marginal land" into loblolly pine production.

They soon wrapped up two big contracts; after which, Conyers moved the crew again, this time across the state to the Suwannee River country on the Gulf coast where they established a new base camp along the edge of a pine plantation. The company had four separate parcels under contract; two of them in a nearby section, totaling several thousand acres, all of it bare ground.

The land had been clear-cut the previous year. There were stumps everywhere and the charred remains of slash piles which had been burned only a few weeks before the planting season.

Next morning, the crew completed the smallest parcels nearest the highway before lunch, then, moved to a more remote tract, miles from any road. Tom and Bill Nelson were now the two lead planters, and traded back and forth. One would lead for awhile, then step aside and let the other one take it. Planting steadily in this fashion, matching tree for tree, they reached the parcel's northern boundary about mid-afternoon, where they noticed several men digging with shovels on an adjoining piece of ground.

Tom and Bill halted for a breather at the boundary. There was no fence but the line was apparent because the abutting forestland was still intact. But it was a different forest, with some other type of pine. A posted sign read:

FINDLEY NATURE RESERVE:
No Hunting!
Violators will be prosecuted

Tom and Bill struck up a conversation. "Hey there. Hello."

"Hello," one of the men said. The two other men also halted work and came over, shovels in hand. They were smiling.

"Always happy for an excuse to knock off," one said.

"We know how that goes. What are you digging?" Bill said. "Looks like a ditch."

"No. Not a ditch. We're making a fire line." The man leaned on his shovel. Tom and Bill leaned on their hoedads.

"You expecting a wildfire?"

"No. No. We're getting ready to stage a burn later this winter. Prescribed fire. We burn this ground every four to five years." The other man approached.

"Hello," he said, extended his arm to shake hands. "Name's Will Hatcher."

"Pleased to meet you. What kind of pines are those?"

"This section," the man said, indicating the forest behind him, "is the largest remaining stand of longleaf in Dixie County. And over there in the swamp along Black Creek – you can't see it from here – is some fine bald cypress. Only a remnant."

"Longleaf pine?"

"That's right. It used to be the main timber tree hereabouts. Originally. By that I mean at the time of white settlement. In those days, longleaf was the dominant softwood species on the coastal plain. Longleaf forests extended over thousands of square miles."

"Is that so?"

"Yes. Given how little of it survives, it's hard to appreciate how vast that original forest was. At one time, it covered ninety million acres and was continuous from Virginia to Florida and as far west as Texas. Today, only small fragments remain. Like this one." He motioned with an arm. "This isolated stand behind us is a part of the two percent that remains."

"What happened to the rest of it?"

There was a silence.

"You men are not from around here, are you?"

"No. I'm from Colorado," said Tom.

"I'm from Wisconsin," said Bill. Three more treeplanters now arrived and joined the group.

"The longleaf forests were creamed off," said Will. "Logged around the turn of the century. Today it's the most endangered forest on the continent."

"You mean in all of North America? That hardly seems possible."

"Well, I hate to disappoint you, but it's true." The men who were with him nodded assent. "Until last year the land you are now planting was also longleaf; originally part of the same stand – but, unfortunately, in the wrong ownership. You may have noticed from the stumps that the site included some very large trees. Surviving longleaf pines of this size are quite rare today and for this reason are extremely important to the red-cockaded woodpecker. The bird used to be common, but with the sharp reduction in habitat, the species is barely hanging on. The birds need large old trees in which to excavate their nests."

The other man interrupted, "There was one hell of a fight to stop the sale, let me tell you, but unfortunately we lost in court."

"That's right. Will here is a biologist from the Tall Timbers Research Station up in Tallahassee. I'm Richard Doolittle, with the Nature Conservancy. At your service." He put his arm around the shoulder of a shorter man standing beside him. "And this fine fellow is Mr. Bo Findley. He's the owner and chief steward of this boot-strap operation.

"Pleased to make your acquaintance."

"Well I was wondering," Tom said, "and uh ... I think maybe you've explained why the pine stands around here look like corn fields, in neat rows."

"The loblolly plantations are not true forests. They don't have the compositional or genetic diversity. They are fiber farms, pure and simple. Agro-business by another name. Loblolly is not even native to these parts. It's a transplant species from up north."

"But if longleaf is superior why are we planting loblolly?"

"For one reason only. Loblolly grows faster. The timber companies get a quicker return on their investment."

"So, why do you burn the stand? You said, every four or five years."

Will paused and gave his colleague an inquiring look. "Richard, want to take a crack at that one? No? Come on, help me out here."

"Help? Why?" Richard said. "You're doing fine."

"OK. OK. Nominated by default. Well, guys, it's a long story. If you have an hour I'd be happy to run through it."

"What we've got is ten minutes," Bill Nelson said. Someone laughed. More treeplanters had gathered around.

"We're on break, y'see."

"OK, well, I'll try my best. Start with this. Gotta start somewhere." He swept his arm around. "Just to look around here you would not know at a glance, or even guess, that fire is and has always been the most important factor in this ecosystem."

"What! Fire? Really?"

"Yes. Most people think Florida is a wet place. And it's true. We are sub-tropical. We have many lakes and swamps and rivers. We also get a lot of rain – anyway, most years. Here in Florida the summer is our wet season, just opposite out West. But with the sandy soils that we have here things dry out fast. Real fast. Within a day of rain we can have wildfire. And this is also true of much of the coastal plain. In the days before the white man in his great wisdom began to suppress wildfire, lightning-caused fire was the dominant factor on this landscape. A site like this would have burned, oh, every one to three years..."

"Wow. That often?"

"Yes. There is no place in North America with a more frequent fire interval. You see, a single dry lightning storm can produce hundreds of strikes. And every one of them is a potential source of ignition. Once started, a lightning caused fire would easily spread over tens of thousands, even hundreds of thousands, of acres – until it reached a river or some other natural barrier. But in those days the fires tended to stay on the ground and they were almost always beneficial. Longleaf pine tolerates this type of fire extremely well. In fact, it and the other plants native to this area actually need low intensity fire to regenerate. Each native species has its own unique adaptations. The end result was an open and park-like forest – a wonderful place – with wire-grass and many different kinds of herbs and forbs dominating the ground layer. And the key element, the thing that maintained the entire system, was wildfire. The native ground cover provided the seeds and food for the many kinds of birds and animals that were part of the longleaf community. The Native Americans who lived here for who knows how long understood all of this, and they made use of fire. But for some reason our European ancestors were convinced that fire was bad, probably because, and this is my opinion, they were afraid of it. After about 1910 wildfire was effectively suppressed in Florida, and the result is what you see. Today, the entire ecosystem has collapsed."

"Collapsed?" one of the planters said, plainly shocked. "You must be kidding. Florida is so green, man, and wet..."

"Yes, but it's an illusion. The average person has no idea of the magnitude of the calamity we are now facing, because of how radically we have altered this landscape."

"Will's right," his colleague added. "Most people do not understand the ecology, especially the role fire played, even local residents, people who've lived in Florida all of their lives."

Will continued, "You see, when we excluded fire, plant succession began to drive the system. Brush and hardwoods from nearby hammock communities invaded the pine stands, moved in and took over. This happened very quickly, within as little as fifteen years. This is why today you see so much palmetto. The native plants and animals were driven out. Yet, and I cannot emphasize this enough, the altered structure and species composition does not change the fact that this ground wants to burn. Indeed it *will* burn. It's not a matter of if, only when. And when it does our worst fears about fire come true in a self-fulfilling prophecy."

A voice interrupted. "So what's the answer?"

"I'm getting to that. With the altered conditions, wildfires now burn hotter, with longer flame lengths, and they tend to be much more destructive. Yet, even so, and this is something many people find hard to understand and even harder to accept, it is still best to let wildfires burn when human structures are not threatened. Because in this kind of ecosystem the outcome of *any* fire will favor the most fire resistant species, the very species that are in the most trouble."

Richard chimed in, "Yes, and what we are trying to do here, in this reserve, and in a number of other places, is simply to hang onto the pieces, the various components of the original ecosystem – which are in danger of being lost – in the belief that one day people will come to their senses, and recognize the wisdom of returning to Nature's original blueprint. Whether or not restoration is even possible in Florida at this late date we can't say. No one knows. But we prefer to think positively about it. As somebody once said, negative thinking has no survival value..."

THIRTEEN

Two days later, droughty weather compelled Ed Conyers to halt the planting. Pine seedlings have very tight requirements and just cannot make it when air and soil moisture drop too low. For a week the crew was reduced to idle boredom. They passed the long hours reading or playing card games. Some tossed a football around, others a frisbee. Still others just lolled about camp with idle hands in their pockets. Frustration became the order of things. There was nothing to do but wait – and pray for rain.

Leave it to working-men to gripe. Even when things go right men will find something, a boot that does not fit properly, indigestion, aching muscles, a splinter under a finger nail, issues with wives or girl friends, high taxes – if not one thing, then another. A certain amount of complaining is actually a healthy thing as it affords men a safety valve, a chance to vent and blow off steam. However, when down time is prolonged the valve works in reverse. Delay becomes corrosive and breeds every kind of trouble for a crew boss. So it happened with Conyers' crew.

In mid-January the work resumed after a lightning storm brought heavy rain. But the damage was done.

A boss walks a fine line between the respect of his men and their contempt, and the balance can shift with astonishing swiftness. Worse, once respect has been lost it is very difficult to recoup. Conyers was not responsible for the dry weather, but some of the crew resented him anyway.

There was grumbling in camp.

A number of the planters supported large families back home, and because of the work stoppage had missed home or land payments. Dire financial straits can drive even good men to do things they would never otherwise contemplate. Trouble was brewing.

Conyers was sympathetic. He knew some of the men were in financial trouble, and he tried to commiserate. "Boys, I know what you're going through. I've been there myself." He described his own personal troubles back in Arkansas. Yet, Conyers was powerless to deliver the necessary rain. That evening the continuing dry conditions compelled him to announce yet another shutdown for the next day. He tried to placate them with the latest weather report. A big winter storm was working its way north through the Keys and was expected to dump heavy rain on central Florida within twenty-four hours. The boss was reasonably certain they would be back to work within a day or, at most, two. But he could not promise them anything. There was no way to be sure.

The crew did not take the announcement well. There was muttering. The tension in camp was palpable.

Tom never knew how or where the trouble started. Shortly after nightfall there was a loud commotion in the kitchen area. By the time he got there Jamie Watters had Sid Ferman pinned against a pickup. Others were standing around.

"Asshole!" Jamie screamed.

"What's going on?"

"I caught him."

"Ain't true," said Ferman.

"You know it is, you dirty thief. He was robbing the kitty. Stealing."

Now, Conyers arrived. "Is that right, Sid?"

"Hell no! I didn't take nothing. I swear it."

"You did. You stole from us! I saw him."

Later, Conyers realized he should have intervened immediately. But hindsight is 20-20. For some reason he hesitated to separate the two men, not for long, only for a few seconds but they were seconds he later wished he could take back.

Suddenly Ferman came up with his knee and caught Watters in the crotch, hard, lifting him off his feet. As Watters doubled over Ferman grabbed him by the shirt and an arm and flung him head first into the camp kitchen. Pots and food went flying as Watters fell through the portable table, loaded with plates and other kitchen-wares.

Everyone stood by flatfooted as Ferman pounded Watters with a cast iron frying pan. Even Conyers was too shocked to move.

Blood was streaming down Watters' face as Ferman pummeled him, again and again, until finally Bill Nelson pulled him off.

"Stop it, man!"

But now Watters twisted free and lashed out with a kitchen knife. Bill staggered back, his face a bloody mess. It happened so fast. Watters had cut Nelson from forehead to chin. Nelson was incredulous as he wiped blood from his eyes. Watters came at him again, evidently so blinded by blood or rage that he didn't know who he was attacking. He lunged to kill.

Bill evaded the thrust, grabbed the man's forearm and wrenched it around with such force he snapped the arm above the elbow. There was an eerie crack, then, a ghastly scream as the knife fell away. Another second and it was over. Bill picked up the blade and threw it into the brush with obvious disgust, then staggered off with Linda to try to staunch his bleeding face. Tallie and several others attended to Watters who was laid out, half-conscious.

Conyers was too angry to speak. They fashioned a crude splint for Jamie's broken arm. After which, there was an accounting.

"What in the hell is the matter with you, boy?" Conyers almost shouted at Watters. He had already fired Ferman and hated losing Jamie as well, because it turned out the boy was right. They found the kitchen money in one of Ferman's pockets. It was hard losing two planters, but what was the alternative? Due to the broken arm Watters was useless anyway. Before the man left for the emergency room Conyers paid him what he owed him. "You're done, Jamie. You won't be planting any more trees this winter. Maybe next year. Good luck to you."

It was the last they saw of either man.

Bill Nelson fared the worst. When he got back from the hospital his face was a swollen pulpy mess, covered with bloodstained bandages. The only upside was that he seemed in strangely good spirits, a fortunate thing because Bill was going to require a skilled plastic surgeon and months, maybe years, of professional care.

How does a crew recover from such an incident? There was no way back. A feeling of gloom descended over camp.

Next evening after dinner, Tom was deeply immersed in Kant as usual when he heard a light tapping on his window. It was Tallie, again.

"Oh. Hey."

"Hi."

"How's it going?"

"OK."

"Just OK?"

"Actually, things are not so great. I need your help, *again*. Do you mind?"

"Of course not. You need a light?"

"No, a ride." She was leaving, she said, and needed a lift to the bus station in Cross City.

"OK. Sure. When? Tomorrow?"

"No. I have to leave tonight." She had a backpack and some other belongings. He also noticed that she had been crying.

"You mean right now? This minute?"

"Look, if it's a problem I..."

"No. No problem. I don't mind. Just ... wondering."

He slid a bookmark into Kant, dimmed the lantern, and made ready to go.

They talked during the drive, their first actual conversation. The fight had powerfully affected her. Tallie blamed herself, especially for what had happened to Bill Nelson's face. She was convinced she had brought the crew bad luck. The only solution was to leave. He disagreed and told her so.

"It wasn't your fault. Stuff happens. Come on. Some of these guys are crazy."

"That poor man will be scarred for life because of me."

"Tallie, you didn't cause the fight. Don't blame yourself."

"Tom, I was responsible for the kitchen money. I kept it in a coffee can and I should have put it away. But I left it out ... almost two hundred dollars." Their eyes met. Tears were streaming down her cheeks.

"Hell," he said.

Now she wept openly, inconsolably. He felt lost. He did not know what to say. He drove in silence. Dark country sped by.

"What about your cousin, Alan?"

She snuffled, "Alan's not my cousin, he's my half-brother. I never should have come down here with him. He thinks he owns me, and..."

There was more to it, apparently. "And what?"

She finally told him. "He started coming onto me. It's why I can't stay. I have to go tonight."

"And Linda?"

"She understands."

He concentrated on the driving. The headlights seemed feeble in the surrounding darkness. "Where will you go? Back to Maine?"

"There's nothing for me there."

"Where then? Parents?"

"I have a friend in California. She wants me to come out there and live with her."

"That's a 3,000 mile trip."

"I have the fare."

But they missed the last bus out of Cross City, and had to spend the night at a motel. He offered to get her a private suite but she refused. She did not want to be alone. They took a room with twin beds.

They were both tired. After saying "good night," Tom turned out the light. He was asleep almost before his head hit the pillow.

FOURTEEN

She awoke gasping for air and knew she was lost. Her skull was in a vise. A mountain of rocks had been piled upon her; and the incalculable mass was slowly crushing the breath out of her, grinding her to dust. The pressure was approaching the outer limits of pain.

Now, the pain increased to another level. Someone inside her cranium was violently pounding with a sledge-hammer. The pounding throb merged into a continuous piercing knife that cut all of the way through her, even to the core of her being. There was no escape. She would have wept but the hot blade had already scorched the tears out of her.

All of this was happening inside her head.

It was extraordinary that in the midst of this waking horror she managed some clarity of thought, because a part of her somehow remained detached. The migraine was like a prism in this respect, a crystal lens that actually heightened her awareness. The pain was a portal through which she went out of herself into a kind of universal witness space that gave her a more clear perspective on her life. Strangely, it was in such moments that she knew herself best.

In this zone there was no room for illusion. The pain was too intense for vain bullshit, too relentless for self-pity. In this respect, the migraine over which she had no control was a perfect metaphor for her own existence. In such moments she saw herself with keen dispassion. She had been running all of her life. She could scarcely recall a time when it had been otherwise. Running from school, from work, from every responsibility and obligation, from the past, from reality, from mature relationships, from herself. Her affairs were a continuing fiasco, little more than stolen kisses, snatched moments of pleasure. They always ended the same way. How many hearts had she broken because of her impulsive faithlessness and her lack of loyalty to anything or anyone? It was as if she lacked a center of gravity. How she loathed the person she had become.

It was paradoxical that in such moments she also felt closest to the ones she truly loved, the family members who had died, and whom she had betrayed by living. She sensed their presence through the pain. During the migraines she relived the head-on collision with the semi, five years before. The car crash still haunted her. She had been in the back seat with her sister Deborah. How many times had she revisited the horrifying moment when the semi veered across the median strip? By rights, she should have died with them. The terrible impact had reduced her father's Volvo to a crumpled mass of twisted steel; only pieces of dismembered bodies had been recovered. That she had come through it with only a bump on the head and some minor cuts and abrasions was one of the imponderables. She had lived, yes, but for what?

Within weeks of the accident the migraines started.

Five years later, she knew that the threshold between life and death was razor-thin; but the only answer she had ever found to the questions that mattered most was a cup of pain.

Now, the migraine intensified, driving out her capacity for rational thought and self-reflection. She descended into unmitigated suffering and absolute despair. How she longed to lose consciousness, yes, even to die. She sensed the proximity of death and longed for it the way a man lost in a desert thirsts for one sip of water. She was no longer frightened. Death had lost its sting. The migraines had long since squeezed the fear out of her. Had she been physically able she would not have hesitated to do herself in, simply to put an end to the pain. It would have been so easy to slip over the final threshold. The peace of death would have brought such exquisite relief. But this too was denied her, for by the time she descended to this level, the pit of the migraine, the abyss, she was invariably too weak to lift her hand from the bed. She was physically incapable of the act of suicide.

Mercifully, later, she never remembered her own suffering. Afterwards, when the veil of pain miraculously lifted, she quickly regained her usual zest for life. Nor did she recall the witness space. It was like a dream that vanishes when one awakens. It had always been like that.

But what's this?

Through the agony she sensed movement near the bed...

Tom was shaving when he heard the moan. He came out of the bathroom with Barbasol on his face and was shocked at the sight of her. She looked so different, so pale. Her normally healthy color had drained from her face, along with most of her beauty. Her skin was tightly drawn over her cheekbones and had a wooden cast. Her breathing was shallow and irregular, her skin hot to the touch. She was flat on her back, almost unable to move. She seemed to have withdrawn into a wraith of herself.

He asked how he might be of assistance. Did she need her pain medication? But she only groaned softly. She was unable to speak. He had no idea what to do.

He opened the drapes to let in the morning light, thinking this might revive her. But she moaned the louder and tried to turn her head away. He realized the light was hurting her, making it worse. So, he closed the curtains tight and also turned out the light in the bathroom. After that, he kept the room as dark as possible.

He was concerned that she might become dehydrated. So, he went out and minutes later returned with a bucket of ice. He sat on the bed and gently spooned small chips into her mouth. She sucked on the ice and seemed grateful to have it. He also fashioned an ice pack, using a towel from the bathroom. For a time she tolerated the ice on her forehead.

He was horrified by the intensity of her suffering. The migraine seemed to him like a descent into hell. He wondered how she could stand it, how anyone could. Her face had taken on an ethereal ghost-like quality, as if she were barely present, as if the pain was driving her out of her body. Judging by her appearance, this level of suffering seemed beyond the limit of human endurance.

He felt useless, so frustrated by his inability to help that for a time he paced back and forth, wondering what to do. Eventually, though, he realized there was nothing to be done but wait and keep watch over her. He positioned a chair outside the room by the door, where he sat reading Kant; or tried to, anyway. But he couldn't concentrate. He thumbed the book, staring blankly at the page.

His thoughts also wandered to the recent lesson in the field. Will Hatcher and the man from the Nature Conservancy had tossed a monkey-wrench into his obviously ham-fisted view of things. Treeplanting had suddenly lost its appeal, the qualities that made it feel worthwhile, and now seemed like an exercise in futility. He was in the throes of yet another rude awakening, another episode of disillusionment. He felt like a cog in a giant faceless machine that was spinning madly out of control. The gears were binding up, the tires coming off. Was everything loco? Had the whole world gone haywire?

No, he could not read. Every so often he looked in to see how she was doing. The hours slowly passed.

The migraine lasted fifty hours; two seemingly interminable days.

Early on the third morning she came out of it. Her recovery was remarkably swift. After showering, the transformation back to her usual smiling self was complete. She laughed as she emerged from the bathroom, drying her hair with a towel. Incredibly, she appeared none the worse for her suffering and even made light of it. Her radiance and healthy color had returned.

She was ravenously hungry, and no wonder. So, after she dressed they went out for breakfast. Tallie ate like a horse, then ordered a second meal which she ate with equal gusto and wiped her plate clean.

They returned to the room a little after 9 A.M.. Her bus to New Orleans did not depart until 1:15 that afternoon. She hung the "DO NOT DISTURB" sign on the outside door knob, then closed the door and made certain it was locked. An impish light sparkled in her eyes. Before he could say a word she pushed him backwards onto the bed, climbed on top of him and tenderly anointed his nose, eyes, and lips with kisses.

Slowly, she unbuttoned his shirt.

Three hours later they rose and dressed without a word. He loaded her belongings into the truck and drove her to the station. They sat holding hands in the lobby while they waited in silence. Speech seemed almost a defilement of the emotional space they now shared. She waited until the last moment to board.

"I don't want you to go."

"I have to."

"No you don't."

"Yes. I do."

"I want you to stay."

"I can't."

He kissed her. "Yes you can." She put up no more resistance when he took her by the arm and led her to his truck. They returned to the motel. He paid for another night. This time, he undressed her.

Afterwards, they showered together. Then, he took her out to eat at a decent Italian place. She was lovely in a simple sky-blue blouse and jeans. She wore no make up. Apparently she never did. With her looks she did not need makeup anyway. She had left her shoes in the truck. She said they made her uncomfortable. She preferred going barefoot. She said she felt more "in touch" this way. She liked to sit with her legs curled beneath her.

He called her "twinkle toes" and "peek-a-boo", which made her laugh. While they ate, the subject of her departure eventually arose. She said with finality, "I will leave tomorrow."

"OK."

"What will you tell them when you return?"

"I don't know," he said. "I'll think of something." After a moment he added, "I'll tell them the truth. If they can't handle it, to hell with them. That's their problem."

They changed the subject.

However, the next day the same thing happened. He took her to the station, but somehow they ended up back at the motel between the sheets.

By the following day they gave up the pretense of going to the station. During the next few days the room was an oasis, a respite from every care. The good things in each of them burst forth, as when a many-colored flower garden appears in the desert after a rare torrential rain. They lived fully in the moment, affirming one another.

However, early on the sixth morning they arrived at an impasse. It grew out of a common understanding. A deep bond was now forming, and both of them sensed that if she stayed one more night she would never leave. This was fine with Tom. He wanted her to stay. But she was stubborn, every bit as stubborn as he was. She said there were things that she could not explain and might never be able to tell him. She didn't want to go there. So why start?

"Things? What things?"

She sighed.

He noticed the strain around her eyes and coaxed her so persistently that eventually he dragged it out of her. Some of it, anyway.

The migraines were only part of the problem, she told him. She was in pain most of the time, physical pain. She lived with it. She had been that way for five years, so long that she could not remember anything different, what living had been like before. She used a scale of one to five to describe the intensity of the moment. A "one" was a good day, a day free from pain. She had such days but they were the rare exception. A "four" approached the outer limits, though she reserved a "five" for the migraines. "Twos" and "threes" were the general rule.

The doctors had no idea what was the matter with her. They had done every kind of diagnostic test. You name it she had tried it. They had palpated her and stuck and prodded and x-rayed her. She had endured numerous kinds of treatment, including exploratory surgery, even three years of psychotherapy to no avail. She had used various types of pain medication which did help for awhile, until she became addicted, and this she could not stand. She hated being an addict even worse than the pain. In the end she had voluntarily gone cold turkey; and was determined never to backslide. She was presently medication free and would stay that way.

She had her own theory about it. She said she was "a freak." She called herself "a marvel of medical science." She was "a medical emergency waiting to happen." She thought her brain was wired differently. She was too sensitive, she said. Her pain threshold was too low. Things that were normal to other people were painful to her, especially bright light and noise. This explained the sunglasses.

There was no cure. Anyway, she had never found one and some time back she had given up looking. She lived without hope. Something inside of her was broken, some deep part that could not be fixed. Simple as that.

He felt the bite of her sarcasm. It seemed so out of character. She was like a wounded bird, innocent and accepting, deeply tragic, without a trace of guile or bitterness. She was blind to her own good qualities and her own beauty. He was astonished, under the circumstances, that she was able to maintain a positive outlook most of the time. Her resilience amazed him, yet, he was appalled by the toll.

About noon on the sixth day they returned to the station. This time – he knew – she was really going to go. She had made up her mind to leave, and he respected her decision. He told her so. He would not stand in her way. All the same, she waited until the last second to board the bus.

Her eyes filled with tears. "Oh, Tom," she said, "what's to become of me?" Then she stepped up into the Greyhound and passed out of his life.

BOOK TWO

FIFTEEN

Jacques St Clair was enjoying his second cup of coffee, and took a deep breath of mountain air. Bracing! A spray of morning sunlight was filtering through the nearby conifers. Songbirds were singing their hearts out, and back in the forest a woodpecker was busily announcing his existence to the world, as he excavated a nest.

Jacques loved the woods, and was especially pleased on this fine day. He had a feeling that fortune was about to smile upon him. Finally.

The previous afternoon, he had finished moving his heavy equipment from Cameron Pass to the new job site at Bowen Gulch; and within a week or so, the woods would be dry enough to start what promised to be the most profitable timber sale of his career.

Like other operators, Jacques preferred logging old growth. No bones about it. From the standpoint of value there was no comparison with second-growth timber. Old growth was like money in the bank. Large diameter, straight-grained, centuries old trees always brought a premium price at the mill. Top dollar. The timber sale volume was supposedly ten million board feet, which was extremely auspicious. Of course, this was on paper. Jacques knew from experience not to trust Forest Service estimates, which were often overblown.

He would believe the estimate when he saw the timber with his own eyes, and that would be soon. Early next week he would walk the entire sale, every unit, with Bill Noonan, the district's chief timber sale officer, who would also give him the final nod to start building the skid trails. Jacques had already cruised the lower units above the main landing which were now snow-free, and liked what he saw.

St. Clair enjoyed operating heavy equipment, though at present he had a couple of issues. One of the steel plates on the left track of his dozer had developed a hairline crack and needed a welding job. Skidder number two also needed servicing. But such problems were the rule in the trade and he took them in stride. Over a career spanning two decades Jacques had scouted jobs up and down the Rocky Mountains, from Montana to New Mexico, and whatever he lacked in panache he made up for with practiced skill and steely determination. He even drew a measure of pride vying with conditions beyond his control, fog, snow, cold, rain. The elements did not overly concern him. However, deadlines, shrinking margins and bureaucratic bullshit were something else again.

Normally, maintenance was Francis Delacour's responsibility, but the boss had just exempted his foreman from some of his duties because his wife, Rosemary, was nearly nine months pregnant with their first child. Francis had already moved Rose to a hotel in Granby to keep her close, and intended to be at her side for the delivery. Jacques thought Rose was already near enough to term to cut his foreman some vacation time. The previous day, he had told Francis to drop everything.

"Stow it. I want you out of here. Rosemary needs you."

"But boss," Francis had objected, "There's stuff to be done. The skidder – "

"Don't worry about it. I'll take care of the skidder. You need to be with your wife. Just go! Get the hell out of here."

"Thanks, boss." Francis strode off toward his truck wiping his greasy hands on his jeans, but turned around again beaming as he back-stepped. "Could be as soon as this weekend. Hah!"

"Give Rose my best, Francis. I will expect you next Monday, bright and early. I'll need your help with the cat. The tread will keep until then..."

"Sure, boss."

"Hey, is it a boy or a girl?"

"A girl. Lucy."

With a laugh Jacques waved him off.

The boss strolled to his pickup. There were errands to run. He needed to stop by the branch bank in Granby to transfer sufficient funds into his business account to meet his payroll and he also planned to check on his crew. By now, all of the men should have moved to the Forest Service campground he had reserved in the Kawuneechee Valley for their exclusive use. The old camp had not been used for years and was conveniently located about three miles from the job site. It was a scenic spot, not far from the headwaters of the Colorado River, with a terrific view of Rocky Mountain National Park.

Most of his loggers would probably scatter for the extended weekend; but Shorty, one of his oldest hands, who was a bachelor, had agreed to stay behind and keep watch over the new camp.

Jacques was more concerned than usual about security because the Bowen Gulch sale had stirred up a hornet's nest of controversy. Judging from the press, feelings were running high across the state. There had already been two demonstrations at the district office. At one, the damn tree-huggers had occupied the building in a failed attempt to shut down the office. There had been a minor scuffle and several arrests. There had also been one case of vandalism, "ecotage" they called it, about which Jacques was outraged. The incident was a case of terrorism pure and simple. Jacques viewed property destruction as a frontal assault on his way of life, yes, and everything he stood for. To be sure, the operator, a long-time acquaintance, had also been partly responsible because the man had foolishly left his equipment unattended.

There had been reports, actually nothing definite, mostly scuttlebutt via the grapevine, that environmental nut-jobs were threatening bigger actions if the sale went ahead.

Of course, one could never tell with such rumors. Probably it was all smoke, so much bullshit. Talk was cheap. But there was no point taking unnecessary chances. It was why Jacques had parked his twenty-foot silver Gulf Stream camper on site, beside his office trailer, where he could keep a close eye on his dozer and skidders.

At the moment, Jacques was actually more concerned about a different kind of trouble closer to home. Shortmanned, as usual, he had just hired two new cutters, Bobby Lighthorse and George Kowalsky, both experienced timber fallers. The men had seemed OK; however, one of them had already tangled with Wolfe Withers. There had been a fistfight; and possibly much worse had narrowly been averted when Francis stepped between the two and pried them apart. It had been a near thing.

Jacques hated this sort of trouble and did his best to head it off. Prevention was always the best approach. He was not certain who the instigator was but he suspected Wolfe, one of the most disagreeable and ill-tempered men he had ever met. The man could cast a shadow on a cloudy day. Jacques recalled the pent-up rage he had sensed boiling just below the surface, and wondered if hiring an ex-con straight out of prison had been a mistake. None of the other cutters would have anything to do with the man. They shunned him. But Jacques believed strongly that everyone deserves a second chance, period, no matter what a man might have done. The bottom line was that Wolfe was an experienced timber faller and Jacques was short-handed, even with the new hires. He needed every saw and could not afford to lose even one.

Better keep an eye on him.

SIXTEEN

There was much to like about Bobby Lighthorse. He was detached though engaged, somewhat of a loner yet good company. He had a quick sense of humor, but also the steady determination of a man with a purpose, as if he were on a mission. There was no mistaking that he was a serious customer, an impression strengthened by the recent blow-up with Wolfe.

Now, as Tom rapped on Bobby's camper door he heard laughter within, and the familiar voice of Dan Rather.

"Yo. Come in."

Bobby was in the breakfast nook, seated across from another newly hired cutter with whom Tom was not acquainted. His name was George Kowalsky, "Pissant" for short.

"Did you hear? We start next week."

"Yeah, I heard. You know Pissant?"

"Not really. We..."

"Wait," Bobby said. "Piss, turn that off."

"What is it? The nightly news?"

"Nah. H.W. Bush just finished a press conference. Same old shit."

The other man had reached for the TV but instead of turning the volume down accidentally turned it up. As he fumbled with the knob the tattooed bust of a nude woman on his right biceps came alive. Tom was amused.

"Piss, show him Marilyn."

When Pissant double-flexed his huge arm the generous bosom inflated in real time. The big breasts heaved and the cleavage deepened. The nipples beckoned. Bobby was smiling. Tom was speechless.

"Not b-b-b-bad, huh?" said George.

"No. Not bad," Tom conceded. "For a tattoo." Under the graven image a name was inscribed. "Who's Marilyn?"

"That's his latest," Bobby said. "Right, Piss?"

"That's wight." The man's chest swelled.

"But what if you change your mind?" Tom said. Obsolescence seemed an obvious concern. But Pissant gave no sign that he understood. Tom tried again. "What I mean is, what if you and Marilyn should break up for some reason?"

"Oh he don't care about that," Bobby said. "Show him Louise."

"Oh, yeah, Loueeth. Wight heah." The big man moved the slab of his other arm into view. It too was a montage. Several female visages were prominently displayed.

"So which is Louise?"

George pointed. "D-d-dis heah's Loueeth. Shee's my ex," the man stammered.

"George has more ex's than Carter has pills," Bobby said. "But Louise looks more like a fish."

"Yeth," said Pissant. "Shee hath thum theeweus ithews."

"Piss, show Tom your medals from Nam."

"Oh sure," said Pissant. With his big hand he stretched the collar of his cotton t-shirt down and to the right, beyond the elastic flex limit. In the process of half-ripping the shirt he exposed a patch of hairless chest below his collarbone distinguished by a row of rainbow-colored bars, bright red and blue ribbons, plus a silver medallion – all indelibly inscribed. Pissant tucked his chin to view his own glory.

"Very impressive carnage I must say. So, what's this other insignia on your arm?"

"It's my f-f-former unit," Pissant said. "The 101st Airborne."

"Wait 'til you see the rest of him," Bobby said.

"I can imagine."

"Nope. I doubt if you can. Show him, Piss." George was only too happy to oblige and peeled off his shirt, revealing more tattoos. The torso was covered front to back, stem to stern. The man was a walking art gallery. Of what remained the question. Tom's mouth fell open.

"Jumping Geronimo! He's an illustrated man."

Pissant glowed with pride.

"Some guys wear their heart on their sleeve," Bobby said. "George prefers inscription."

"You wanna thee my Chineeth dwagon? It's weel cool, a minature dwagon, with fiewy bweath."

"Sure," Tom said. "But later, OK?" He had come to visit with Bobby. "The boss was just in camp checking on Withers," he said. "I noticed the two of you parked your campers as far away from one another as possible."

"That was Francis' idea."

"What set him off?"

"Wolfe has a hard-on for trouble."

"The way he went after you. The guy is violent." Tom recalled his sole encounter with the man and the cold chill that had passed through him.

"On some level the sick bastard knows I've come for him."

A long pause. "What do you mean?"

"Exactly what I said."

"But..."

"Look, I've been tracking Withers ever since he got out of prison. It took me awhile but I finally found him. They don't call me 'Lobo' for nothing."

"As in wolf..."

"It takes one to catch one."

"But why?"

"Justice, brother. Or rather, because there is none."

There was a sketchy tale about a mate who stumbled onto a drug deal going down in an alley off Larimer Street; wrong place, wrong time.

"Happened 'bout a year and a half ago," Bobby said. "Withers slit Johnny's throat."

"We heard Wolfe was in Canon City, up for murder."

"You heard wrong. He walked. They sent him up for possession but he was paroled after only a year."

"That sucks."

"Yeah, it does."

"So, you are here for revenge?"

"You think like a white man."

Tom noticed the mushrooms lined up on the counter. They were so small. He examined one but was at a loss to identify it.

"I've never seen this kind."

"You won't find these in any mycology book. They are exceedingly rare. You have to know where to look, and they only produce fruiting bodies after a really wet spring..."

"Like this year."

"Right. Maybe one year in twenty. But they are well known to my people."

"What's your tribe?"

"Arapaho. And some Paiute and Shoshone."

"I g-got an eighth of Chewokee in me," said Pissant. "B-b-but I geth it ain't enough. Cauth I never heard about no m-m-muchrooms."

"What are we going to do with this crazy Pollack?" Bobby said.

"Scalp him."

"Make a lampshade out of his silly hide..."

"Yeah, light him up."

"I ain't no Pollack," Pissant objected. "I'm from D-D-Dallas."

"I don't suppose these are the culinary variety."

Bobby laughed. "You suppose right." He picked up one of the caps. "You want to know why I'm here, I'll tell you. I'm after justice, Arapaho-style, and these little muchachos are going to deliver it. These are my allies." He paused. "Oh, I would effing enjoy killing that son of a bitch, doing to Wolfe what he did to Johnny. Withers is one seriously messed up hombre. But that would only perpetuate the cycle of violence, and also create bad karma for me, which I do not need. I have enough of my own. No, I'm here to break the chain. To cure him, if possible..."

"And if it's *not* possible?"

"Then, I'll lead him to the edge, and smile as he goes over."

"Meaning?"

"We have a saying.... Give an evil man sufficient rope and he will hang himself, every time."

"Hard to believe these itty-bitty mushrooms could do that. I mean, they are so tiny."

"Size is relative."

"What do you call them?"

"We call them ... well, the English equivalent would be 'truth serum' or 'truth medicine'. Don't be fooled by their size, amigo. Trust me, the life force, or spirit power, in these mushrooms is strong beyond reckoning. Strong enough to reach down deep into a man and bring out whatever he holds in the cave of his heart, whatever may be hidden there, good or evil, darkness or light."

"Or maybe some of both?"

"Let's hope, for his sake, that you are right. An honest man has nothing to fear from these mushrooms." He smiled. "I intend to partake of them, myself." After a long silence, he added, "But I cannot do it alone, what *must* be done. I am going to need some help."

"You want *my* help?"

Bobby nodded.

SEVENTEEN

Judging from the map on the backside of her letter, the ranch was ten to fifteen miles south of Granby, off the highway to Berthoud Pass.

On his way out of town, Tom stopped at a self-service station and was filling up his tank when a pickup appeared on the far side of the island. The driver's face looked familiar. Tom was certain he had seen the man before. But he could not recall when or where. He studied him trying to think. "Excuse me. Say, you look familiar. Don't I know you?"

"Yeah," the stranger said. "You look familiar too." Another moment and they both made the connection.

It's the green-eyed stranger!

"Oh hey, you pulled that gorilla off of me."

"Yes! That time at the Nugget."

"That son of a bitch almost killed me. Did you know he broke my nose?"

"No I didn't."

The other man fingered his nose gingerly. "You can see it's still red, and crooked from where he stomped on me."

"Yeah, I can tell. Sorry about that."

"It still hurts."

"Shorty Dibbs."

"That guy ought to be on a chain. He's an animal."

"It was a mistake to taunt him like that."

"He's dangerous."

"Actually, he's not that bad, he's OK. Except when he drinks. Shorty can't handle alcohol."

"Yeah, well I don't respect loggers. I've walked across too many clearcuts."

"Never got your name."

"Richard," the man said. "Richard Peters. Pinecone for short." He smiled as he extended an arm.

"Name's Tom Lacey. Happy to know you." The two men gripped hands.

"I never had a chance to thank you. I'm grateful. You really saved my neck."

"Forget it. Are you with the Sierra Club?"

"Hell no. Earth First."

"Earth First? Never heard of it. What is that?"

Tom listened while the man talked about radical environmentalism, first in general terms, then, about a place called Bowen Gulch. He was passionate. "There's going to be a big timber sale over there."

"I know."

"You do? How?"

"I'm part of the crew that going to log it."

"Really? Have you seen it?"

"No. Not yet."

Pinecone explained that Bowen Gulch had originally been part of the proposed Never Summer Wilderness. "Only they excluded it because of heavy lobbying from the timber industry. The politicians adjusted the line before the wilderness bill made it out of committee. They shafted us again. Happens every time."

"So what do you intend to do about it?"

"Stop the sale. Whatever it takes. That's why I'm here." A wry smile appeared on his face. "I've been busy, doing some leg work."

Tom knew better than to ask. "Well, you better get a move on. You're running out of time. We are supposed to start logging up there, next week. For sure by next Thursday. Maybe as soon as next Wednesday morning."

"Oh yeah?"

"Yeah."

"Are you sure? How do you know?"

"I got it straight from the operator, Jacques St. Clair. He's my boss."

"Damn." Pinecone was plainly distressed. There was a pause. "Tom, what are you doing, right now? This minute."

"I'm headed out of town. It's my day off."

"Before you go why don't you come with me. To a meeting..."

Tom did not like the sound if it. He hated meetings, which he always found boring, like being in class. "Nah. I'm due down the road."

"There will be people from all over the state, including some really interesting folks I'd like you to meet. It won't take long. Come with me. It will probably only last about an hour."

"Well, I..."

"Hey, it's not far. It's only a few minutes away. Follow me. I'm headed over there, right now. The meeting's supposed to start at eleven A.M."

"Hell. OK. Why not?"

He followed Pinecone to a derelict house on the outskirts of Granby. The overgrown yard was wall-to-wall tents and campers. The driveway was a parking lot back to back with vehicles. Tom finally found a parking space down the block. Pinecone was waiting for him on the driveway, and escorted him into the backyard where the meeting would take place. People were talking and milling around. He and Pinecone found two vacant seats at a large picnic table, one of several in the yard.

It turned out that Pinecone was a biologist, and also owned an advanced degree in Natural Resources Management. After picking up his masters he had joined the Forest Service where he worked for a half-dozen years, but had become increasingly disillusioned. Pinecone explained that he was the biologist on a team that prepared timber sales and had watched in dismay and with mounting anger as agency staffers paid lip service to ecological principles while they proceeded to push logging roads into one roadless area after another. His persistent attempts to change the extractive policy from within the agency were frustrated at every turn. When he could not take any more, he finally cut bait.

"What's the purpose of the meeting?"

"A strategy session. You'll see."

All of the tables were now fully occupied, and several people began bringing folding chairs from the house and setting them up. From the backyard, Tom could see what appeared to be a traffic jam out on the street as numerous drivers searched for parking places.

A steady stream of people now flooded into the yard. Most of the folks were young, but not all. Some were middle-aged. Tom noticed a few elderly couples. One old man sported a t-shirt with "KEEP IT WILD!" emblazoned on the front in bold green letters. An elderly man and a woman each had posters. One read, "STUMPS DON'T LIE!" The other said, "SAVE YOUR MOTHER!"

The atmosphere was friendly but from the tone of the conversations taking place all around, emotions were pitched. Tom overheard a woman say, "And when they finish with Bowen Gulch it will look like every other place in the state they've trashed."

"Exactly right," said the man sitting beside her.

"The only thing the bastards understand is 'take the money and run!'"

The backyard was filling up and still they came. The crowd already numbered over a hundred people.

Now, there was a palpable feeling of anticipation. Two individuals up in front unfurled a large banner that read: SAVE BOWEN GULCH!

Finally, a woman rose with a microphone. "Can everyone hear me?" she said. "How about you in the back?" There was a loud squawk, then a commotion as several people hustled to adjust the sound system. "There. That's better. Thank you, Earl, and Chuck. Can everyone hear me, now?" Heads nodded in the rear.

After introducing herself, the woman welcomed everyone and thanked them all for coming. Tom did not catch her last name. Leslie something. She was, she said, the spokesperson for a broad coalition calling itself the Ancient Forest Rescue, an umbrella group made up of local, state and national environmental organizations. Leslie began to introduce the steering committee. As she named each one they rose and said a few words about their organization and how they were contributing. The committee had about a dozen members. Some of the national environmental groups were represented, including the Sierra Club, the Wilderness Society, and Audubon. Leslie herself was with the Sierra Club. Yet, it was evident that the coalition was largely a grassroots effort. Nearly all of the members were Colorado residents, with Denver and Boulder well represented. But folks had also come from many other parts of the state; and from a show of hands there were even a few from neighboring Wyoming and Utah.

"In a minute," Leslie said, "I will turn the meeting over to the man who started it all, Dr. Mickey Newsome, and, folks, let's hear it for Mickey, we owe him and his coworkers a huge debt of thanks." There was an enthusiastic round of applause, which went on for awhile, then, slowly subsided. "However, before we hear from Mickey I need to make an announcement. I have some breaking news that is very positive and exciting. Each and every day we are growing in numbers and support and gaining momentum around the state; and there has just been a major development that shows we are close to critical mass. In case you did not hear, earlier this week, the Boulder City Council passed a resolution officially endorsing a statewide boycott of wood products made by Western-Pacific. The resolution will go into effect *if* the company proceeds with the Bowen Gulch timber sale." More cheers. "As you probably know, the Boulder County Commissioners have already endorsed a boycott. So, this latest decision shows our growing strength and is a wake-up call for W-P. We have begun to hit them where they live – in the pocketbook. For the first time, W-P has an obvious financial incentive to cooperate, and abandon the sale. So, let's hear it for the Boulder City Council." There was another enthusiastic round of applause.

The woman said a few more words, then passed the mike to Dr. Newsome, a math professor at the University of Colorado. Tom was surprised by his appearance. Newsome had long hair and looked more like a flower child than a professor. Yet, only weeks before, Newsome and several others had launched the Bowen Gulch campaign by organizing a 24-hour vigil on the steps of the state capitol in Denver. Tom already knew the basic story from a flier someone had passed around. The demo at the capitol had been billed as a memorial service for a 600-year-old spruce tree, a huge log that Newsome brought in for the occasion in the back of a flatbed truck. Someone had attached markers to the tree rings, one of which supposedly showed that the spruce was already 100 years old when Columbus discovered the New World. That night, eight activists, all now legendary, had slept on the capitol steps. The embryonic event had attracted a surprising amount of media coverage around the state.

After the demo Newsome and company took the campaign on the road. Starting from Denver, for the next four weeks he and a few others trucked the huge spruce log from one end of Colorado to the other, visiting just about every town of any consequence in the state. The roadshow turned out to be a brilliant strategy from a publicity standpoint, because it generated almost daily news coverage. This kept the controversy over Bowen Gulch alive and in the public eye. The press increasingly reported Newsome's travels. As the media coverage steadily expanded it came to include educational sound-bytes, even occasional interviews with distinguished scientists who explained the ecological need to preserve the last remaining stands of ancient forest.

When the professor stood up to speak he was drowned out by applause and cheers. He had to wait for at least a full minute. When the crowd finally quieted down the professor thanked everyone for coming. Then, he tossed gasoline on the fire: "ninety-five percent!" he roared. "I'm a math professor, and I know the numbers. That's how much of the old growth they've already logged. Folks, it adds up to 70,000 acres a year nationwide that we are losing and I'm mad as hell about it! I'm here to tell you I'm not going to take it anymore!" His booming voice cut to the chase. "Are you with me?"

A wave of sympathy swept through the crowd and they shouted back, "Yes!" and "We're mad too, madder than hell!"

Newsome went on. "We in the United States have a long history of liquidating natural resources, especially ancient forests. We are very very good at it. We have become efficient liquidators and the reason is not hard to understand. We've had plenty of experience! 200 years of practicing cut and run. In fact, we have succeeded so well that we are now down to the last 5%. Meanwhile, by contrast, our scientists only began studying ancient forests about twenty years ago; and because ancient forests represent a vast storehouse of knowledge, hardly any of which we have figured out yet, you can see why the last remaining 5% is absolutely critical. If we do not have the sense to stop, now, if we mindlessly liquidate the small amount of ancient forest that remains it will be the equivalent of destroying the library before we've had a chance to read the books. And folks we mustn't, we can't allow that to happen!"

He paused during another round of applause. At length, he raised his palm to quiet the audience. "So, my friends,

I'm sure I don't have to tell you, time is short. The truth is, nationwide we have only a few years to turn this thing around. In the case of Bowen Gulch we have only a couple of weeks, possibly even days. So, time is of the essence. What you and I do in the next 48 to 72 hours may prove critical..."

After more cheers, Newsome's tone modulated. "My friends," he said, "we also need to keep in mind that loggers are not our enemy. They are just working people like you and me. This is not an abstraction. I know from personal experience. I know a number of loggers very well and how they think because, heck, I'm one of them. My dad worked in a mill for most of his life. I speak their language and I am convinced they can change. Folks, believe it or not, loggers do have the capacity to change. The Forest Service, we can change them too. Even Western-Pacific. They can change the way they do business. If there's enough of us we can make them change. But here's the deal. It is going to take a strong push to make it happen, and we are the ones who will have to do the pushing. It's up to us. We are the catalyst for change. Are you with me?" Again, deafening applause and cheers.

A few people began to chant, "Save Bowen Gulch! Save Bowen Gulch!" It spread until the whole backyard was rocking together.

Newsome was done. He had come to fire up the crowd and he had succeeded. Now, as he moved to the side he handed the microphone to Pinecone. Tom was surprised. He had been listening so intently that he did not notice Pinecone get up and move to the front of the crowd.

As the chanting continued the two men exchanged words. Then, the professor stepped back and Pinecone waited for quiet. But the crowd was worked up and did not calm down for a couple of minutes.

Pinecone turned out to be a compelling, even inspirational, speaker. He explained the Bowen Gulch timber sale in detail, and spoke about the biology of the place. "What they are proposing," he said, "is one of the highest elevation timber sales ever contemplated in the Rocky Mountains. If the Forest Service timber sale planners have their way, about 640 total acres will be cut, mostly Engelmann Spruce and Subalpine Fir, including the largest trees I have ever seen in Colorado, some of them up to five-feet in diameter. And all of it is higher than 10,000 feet in elevation. One of the sale units actually extends above 11,000 feet. This means they will be logging at timberline, which is totally unacceptable from the standpoint of responsible forestry. At that elevation it is questionable whether forest regeneration can even occur. I've talked with scientists who oppose the sale for this reason alone. They think it is too risky. They tell me the sale amounts to a large-scale experiment. Folks, I agree with the scientists. We should not be playing roulette with our forests." Pinecone had to stop again and wait for the applause to quiet down.

"Beyond the 'repro' question, there are obvious concerns about strip-mining one of the last remaining stands of ancient forest in the Rockies. Folks, this high valley is so special, it has to be seen to be believed. I am not asking you to take my word for it. Later this evening, here at the house, we are going to show a documentary film about Bowen Gulch. All of you are invited. And for those of you who have not yet visited Bowen Gulch I want to encourage you to do so, at the first opportunity. There are free maps on the table, over there, by the door. We have found that when people see the place with their own eyes, see the incredible forest that we are in danger of losing, they come away enthusiastic about our campaign. Seeing is believing. It's also why we have been leading free public tours into Bowen Gulch. We've been doing about one tour a week, and if I am not mistaken another one is scheduled for this coming weekend. Is this correct, Cynthia?"

A woman seated at his left nodded in the affirmative.

"Good. So, those of you who are interested please sign up after the meeting. Cynthia has the list. She is organizing the next group hike. Cynthia, would you please raise your hand so the folks will know who you are? OK. So, see Cynthia. She will fill you in. Thank you, that's about it..."

Pinecone was done and handed the mike to a woman from Denver who talked about a letter-writing campaign in the public schools which, she said, had already generated more than a thousand letters to the governor and the Forest Service, most of which had been written by young school children. She also talked about the next phase of the campaign, direct action; in other words, civil disobedience, should this become necessary. Training sessions in non-violent protest had been happening for several weeks and another was scheduled for the next morning. She emphasized that the training was mandatory for anyone who intended to participate in actions involving the possibility of civil disobedience.

There followed a lively debate about the forms that direct action would take, especially civil disobedience, and about what was and was not acceptable. There was no consensus. Some defended tree spiking, even the destruction of private property, or "ecodefense", while others adamantly opposed such tactics. The discussion became quite heated with strong views on both sides, and continued after the meeting broke up into smaller groups.

EIGHTEEN

The footpath obviously had not been used for years. She picked up the old trail at the edge of the meadow and followed it toward the forest. The meadow was thick with wildflowers and along the way she paused to investigate delicate bluebells, asters and purple lupine. Momentarily she lost the trail in the high grass; but she knew where it led, quickly found it again and went on.

The trail entered the lodgepole forest and began to climb. She knew it was not far to the top. She loved hiking to her favorite spots and to places she had never been. Along the Maine coast she had often wandered alone from the forested headlands down rough unmarked trails to the rocky shore where the breakers roll in from the Atlantic. Parts of the Maine coast were still semi-wild and she had often spent the best part of a day climbing over the craggy rocks and exploring the tidal pools, often without meeting anyone.

The previous winter, while in Florida, after the tree-planters had headed off to work each morning, she would go on long solitary walks through the surrounding Loblolly pine plantations. A few times she wandered deep into the cypress swamps. Unlike the two other women, she never worried about snakes, alligators or losing her way.

The trail now became steeper. She clambered up the last part over bare rock, already feeling the old thrill. A moment later she stood atop the summit of the ridge. Through a break in the trees the heights of Trail Ridge and the Colorado high country dominated the Eastern sky. The snow-capped peaks made her ecstatic. The view from her aunt's house was good but this was even better.

She turned when she heard some honking and watched as a flight of Canadian geese flew over. They were headed north.

After awhile she kicked off her shoes and sat down with her legs crossed, like she always did. She removed her sunglasses and shook her pony-tail back over her shoulder. The indirect sunlight here under the pines was gently dappled and not a problem for her eyes. She had been feeling so much better and was having another good day.

She tossed away her thoughts and opened to the pine-scented woodland and the clear blue sky. She allowed herself to go deep into just being in this old familiar place. For the next few minutes the world and everything was simply grand.

Then, she remembered the cookies in the oven.

Oooops.

She got up and retraced her steps, but she did so without haste. She was still in "explore mode" and paused to study a shooting star here, a spider web there. Nor did she bother to put her shoes back on before inching down the smooth rock surface. One shoe dangled from each hand. She was a barefoot kind of person.

The highway south out of Granby ran straight for several miles across a broad valley. Tom was still grappling with himself about the letter. Her curious choice of words had taken him back to the difficult period after the separation when over the course of many weeks he had reviewed every aspect of the affair. There had been no shortage of time in which to reflect. At the crux of his ferment was the question, had it been real, or, a wild fling that was best forgotten?

During that period he had tipped like a seesaw first one way and then back the other, as he weighed his feelings and, increasingly, his doubts. Emotionally he had been all over the map. But reason ran deep in him and as the months passed, his doubts increased and the rational side of him gradually gained the upper hand, though not without a tug-of-war. As the affair receded further into memory, he began to harbor a deep skepticism about her.

Eventually, he had concluded that the affair was unrelated to his present mode of life; and the passage of time only sharpened this perspective. Past was past and it was best to leave it that way; in other words, behind him. He had moved on. Let it go. Keep moving forward. Best forget about her. Such had been his state of mind, *before* the letter.

The address led to a line of mailboxes. He slowed and turned to the right onto a country lane, which he followed along an irrigation ditch brim full of fast-moving snowmelt. After a turn, the lane crossed a wide pasture. Within a half-mile he came to a wooden gate. He got out and unhooked the latch. The gate was made rigid by a pair of diagonal pole braces and it opened easily. He closed it behind him and continued across a lush pasture. Four quarter-horses and a young colt were grazing the high summer grass. The horses lifted their heads and flicked their ears as he passed. They were curious, yet wary. The frisky colt bolted and pranced off kicking its heels. The others moved away as well but soon lowered their heads, back to browsing.

He topped the rise of a broadly sloping meadow and the log house came in view. It was a handsome two-story gabled affair with a large stone chimney and a bright red metal roof. Several outbuildings flanked the house, including a greenhouse and a barn. There were large cottonwoods around the place and a stand of ponderosa pines at one side. He slowed to a stop. He needed a moment to think.

This time he would keep his feet firmly planted on terra firma. No more living in fantasy land. He would tell her in plain language that he had moved on. From the note he guessed she had done likewise. They would sort things out together. A reunion would be therapeutic for both of them. They would laugh at themselves and at how silly and immature they had been in Florida. Everything would be well.

As he approached the house two black Labrador retrievers sounded the alarm. They circled his truck howling but when he rolled down the window and spoke to them they turned friendly; and when he climbed out the dogs came up and begged for attention. He scratched their heads and squeezed their ears. They nudged him playfully, licked his hand, and followed him up the flagstone walk, their tails wagging like a pair of metronomes.

He had rehearsed what he planned to say, but the moment he saw her all of that went out the window.

She was pulling a tray of chocolate chip cookies out of the oven when she heard the dogs barking out front. A thrill passed through her. She set the tray on the counter, pulled off the mitts, undid her apron and hurried out of the kitchen.

"I see you met Rough and Ready," she said from the porch.

"I almost burned your letter." Even as he said the words he was shocked by his own voice. In Florida, when she mentioned her friend in San Francisco he wondered about the relationship, though he did not ask her about it because the details...well, he was just not that curious. However, it had been six months without so much as a postcard and, dammit, he felt jilted. He couldn't help it.

"Don't be that way," she said. She took him gently by the hand and led him up the steps into the house, then, up the broad stairs to her bedroom.

Three hours later when they came down again, she was barefoot as she set a platter of freshly-baked cookies on a large round oak table in the kitchen. Her eyes were dancing. "Tea? Or milk?"

"Milk."

She pulled down two glasses from the cupboard and set them on the table. He followed her with his eyes. She was borderline awkward, but when she moved a certain way she morphed into something else that was indescribably beautiful. He could not help but stare, watching for the seamless transition, even in the simple act of opening the refrigerator. She was poetry in motion.

She gave him a naughty look as she poured the milk out. Then, she curled up on the chair beside him. They joined hands, their fingers intertwined, and feasted on the cookies.

"My favorite," he said, "chocolate chip."

"I didn't know you were the jealous type," she said, adding an inflection with her eyes.

Suddenly furious, he stared back daggers; but only for a moment. She had acquired a milk mustache that he thought was quite becoming. "Have a look, Bozo," he said, nodding at the mirror on the wall.

She turned to the mirror and let out a shriek. Like an elf she jumped up, grabbed a wet dishrag from the counter and wiped her mouth, then, with a neat backhand lofted it over her shoulder into the sink. Her face was flushed, her freckles turning that weird shade of green. She came up behind him, wrapped her slender arms around his neck and began to roughhouse.

"Hey."

"Hey yourself." He pulled her around until she faced him squarely.

"I wouldn't have blamed you," she said, indicating the letter.

"So, what did you mean by 'back among the living'?"

But she was already pulling him by the arm. "Come, get up, get up and I'll show you the rest of the house. It's way cool." She led him through a kind of atrium with a skylight past an upright piano into the central living area. The furnishings were spare but in excellent taste. The room was commodious, all in stone and wood, a blend of utility, comfort, and beauty. An assortment of prints and framed photos of horses adorned the knotty-pine walls. Navaho rugs partially covered the flagstone floor, thrown this way and that. The south-facing wall was mostly glass. The large windows rose to a vaulted ceiling, which created the impression of tremendous vertical space. The large room felt even larger. The stairway was at one end, and at the other a massive stone hearth, obviously the work of a master stone-mason. The mantle and chimney were impressive.

But the main attraction was the library. The entire north wall was comprised of bookshelves, row after row. Additional stacks had been arranged at right angles. It was easily the largest private collection he had ever seen. He wandered among the shelves, pausing here and there to sample the collection which was wide ranging. The library also showed refined sensibilities. Several shelves were loaded with the classics, including the works of Plato, Josephus, Homer, Hesiod, Virgil, Ovid, Philo, Plutarch, Herodotus, Aristotle, Plotinus, the Greek playwrights, and many others. There was also a large literary section featuring both classic and contemporary authors. On the adjoining shelf was an extensive Native American collection. Considerable space was devoted to science, especially natural history. Several shelves were filled with books about psychology and religion, including Buddhist, Christian, Gnostic, Hindu, and Sufi texts. There was a section devoted to history and politics, and a shelf of how-to books, plus a collection about horses. Finally, there was a shelf of technical volumes and journals on various agricultural and business-related subjects.

She met his inquiring look and said matter-of-factly, "It's my aunt and uncle's library."

"So, what's your surprise?"

She looked thoughtful and seemed about to answer, but just then they heard wheels on gravel. A green International Harvester carry-all suddenly appeared in the driveway. They went out together and, surrounded by wagging retrievers, helped her aunt Mary unload a dozen bags of groceries and other store-bought stuff. Tallie introduced him.

The older woman was all smiles and after greeting him, announced that company was coming for supper. "And they'll be here at six," she said in a tone that implied "Help! I'm running late, as usual."

Mary and Tallie headed into the kitchen. Tom retrieved his bag from his truck, and went upstairs to shower.

When he returned twenty minutes later delicious smells were emanating from the kitchen. With a flourish Mary removed her apron and went up to change. Tallie stayed to finish dinner and fix the salad. She recruited Tom to set the table.

Fifteen minutes later the guests arrived, a pleasant looking older couple. Tallie encouraged them to make themselves at home in the living room where Tom joined them. After introductions Tallie served wine and cheese, then disappeared once again into the kitchen. The man's name was Peter Martin. He was an archaeology professor from the University of Chicago, and was accompanied by his attractive wife, Stephanie.

A moment later Mary came down the stairs adjusting an ear ring and greeted her guests. After an enthusiastic round of hugs and hellos they settled into the business of catching up.

"When did you get back from Paris?"

"Last weekend," said Stephanie.

"Marvelous!" said Mary. "Paris! Wow! I can't wait to hear about it."

"It was my treat," said the professor.

Stephanie asked about the equestrian trophies on the mantle. "In a former life," Mary replied, "I used to compete. Ancient history now. Speaking of which, Peter has been part of an archeological expedition." Tom understood this was for his benefit. "In Egypt."

"Yes, in the southern desert," said Peter.

"Didn't you say you've been working on a book?" Mary asked. "Or did I misunderstand?"

"Actually," Peter said, "it will be an anthology one of these days."

Tallie came in and announced that supper was ready. They carried their drinks to the table where they were immediately refilled. The food was served and what a meal it was, homemade stew with dumplings, a fresh fruit salad, and biscuits straight from the oven, light as air, served with butter and homemade blackberry preserves.

While they ate Mary got her detailed report about Paris. The conversation then lurched briefly into politics, and eventually worked its way around to the other dinner guest.

"Tom's a logger," Tallie said with unmistakable disapproval.

Tom set his fork down and looked at her.

"Really?" said Stephanie.

"You run a chainsaw?"

"Yes."

"Those noisy things scare the daylights out of me," Peter said.

"It isn't for everyone."

"Tell us about it, what you do," Stephanie wanted to know.

Tom briefly described the job, the woods, the boss and crew. When they ran out of questions he turned to Peter, "Sir, what will your book be about?"

"I always forget the name of the site," Mary said.

"Nabta Playa," said Peter. "It's in the middle of nowhere. No roads. No motels. We fly in and basically have to rough it. There is nothing but desert for 100 plus miles around in every direction. It's the most forbidding place I've ever seen, by far. The southern Egyptian desert is a furnace. Daily temperatures, even in winter when conditions are tolerable, often exceed 110 degrees. But eight thousand years ago it was a very different place. In those days Nabta Playa was a grassland and had a temperate climate; and a sizable human population." The professor paused. "But I'm starting to ramble," he said. "Occupational hazard, I'm afraid."

"No, please go on," said Tallie.

"How did you learn about the place?" Tom said.

"Our work at Nabta Playa dates to 1973 when several of my colleagues happened to be passing through the area en route to somewhere else. By chance, they had stopped for a rest or lunch break and discovered potsherds, which are bits of old pottery, one of the best indicators that a site may have archeological significance. The team returned the following year and we've been digging ever since. They recruited me in at the start of the third field season."

"So, what have you found? Anything important?" Mary said.

"Oh yes," Peter replied. He explained how, during the third year, when the team began to investigate some unusual rock outcroppings, they were shocked to discover that the large stones were in fact megaliths. A very long time ago, someone had moved the stones into position. It was clear that the placements were intentional. "There is no longer any doubt," Peter said. "But who, and for what purpose? We still do not know. Later, we found a standing circle reminiscent of Stonehenge, though much older, and numerous other strange stone formations, all of them artificial. We also recovered many artifacts. We've shown that humans occupied Nabta Playa over a period of many thousands of years. The oldest potsherds date to the close of the last ice age. Some of the largest stones also date to the same period."

"Wow! That's really old. But how could you date them?"

"Good question. Some of the stones were buried under eight to ten feet of sediment. They call it 'a playa,' you see, because it's a kind of natural basin. That's what 'playa' means. Long ago, when the place was lush grassland, the basin collected seasonal rains. And it's possible, fortunately, to estimate the rate of deposition with reasonably good accuracy." But now Peter hesitated, as if searching for the precise words. "Believe me ... it has been immensely satisfying to work with talented professionals. Wendorf's team, of which I am proud to be a part, is a great bunch of people. And the work is...very satisfying. It's such a thrill, I can't describe the feeling, just incredible, to uncover literally a forgotten epoch of human history. Still, I must confess, the work at Nabta Playa has also been the most frustrating experience of my career. It is humbling to study an ancient site for ten years and come away with no final answers. In truth, we are no closer to understanding the significance of Nabta Playa, today, than when we first found it. Some of my colleagues think the stones are aligned to stars. But this is still unproven. The stones remain a mystery..."

"But I love mysteries!" said Tallie. She had come from the kitchen to announce dessert.

Mary suggested they retire to the large room. Peter led the way and proceeded to feed the fire already burning in the stone hearth. When they were comfortably seated around the fire, Tallie served coffee and generous helpings of lemon meringue pie, made from scratch. They enjoyed the pie and coffee while Mary regaled them with stories about the old Wimmer spread in southern Colorado where she had been born and raised. The old ranch was located near Westcliffe, in the lee of the Sangre de Cristos Mountains. In her late-forties, Mary was still an attractive woman. Intelligent and charismatic, she had a way of talking that put you at ease. Finally, she said, "I expect Bernard back this evening. Sometime late."

Tallie turned to Tom, "He's my uncle and ... that's your surprise."

"Um..." Tom said.

"Uncle Bernard knew your faculty adviser."

"Leadbetter?" Tom stammered.

NINETEEN

In Florida they had discussed many things. Actually, he did most of the talking, she the listening. She did not enjoy talking about herself. He talked to open her up.

He knew she did not approve of the logging. But once he got rolling, a light appeared in her eyes. She was definitely interested. Once when he paused, she peppered him. "But how did you go from being a philosophy student to operating a chainsaw?"

So, he told her about Carl "Battery Acid" Olsen.

"I had known Carl from before, through my faculty adviser, Nolan Leadbetter, who hired him during my sophomore year at state. Professor Leadbetter was a leading zoologist, and that summer he was trying to complete his grand-opus, a study of the prairie dog in Colorado. The study had been in the works for ten years. But Leadbetter needed an experienced field guide for the final phase of the project. He recruited Carl because "Old Vinegar" had a reputation as the finest guide in the state.

"Wait. I'm confused. You studied zoology? You told me you were into philosophy."

"I'll get to it."

"OK. Tell me about Olsen."

"I want you to meet him. He's a maverick. Way back when, during the 1940s and 1950s, Carl had been a member of the U.S. Geological Survey team charged with the geodetic mapping of the Colorado River basin; after which, for many years, he pursued a career as a big-game hunting guide. At which he was very successful, and much sought after. Carl's trademark was a guaranteed clear shot. He knows the high country better than any man alive, he calls it the "Big Open." Most of his business came through referrals from friends or acquaintances. Oh, he occasionally led large hunting parties, but he preferred a more personal experience, and usually hired out to individual trophy hunters. Carl's retired now, but he keeps busy running a small post and pole operation over in North Park. That's where I had my start in the woods. He's based in Walden.

"Why do you call him 'Old Battery Acid'?"

"Carl's in the habit of speaking his mind and has a sharp tongue. But when you get to know him you discover he's actually very open-minded. Only, he refuses to suffer fools.

In the spring of my senior year, I had taken a job as a managerial assistant (a glorified file clerk) at the Larimer county courthouse in Ft Collins. I needed the income after completing my course-work in philosophy. I'll get to the switch from zoology. I was flat broke after four years of academia.

But I guess I'm just not suited for office work. Truth is, I hated the endless paper shuffling, the gossiping secretaries and my harping supervisor, not to mention the mind-numbing tedium. I was bored to tears. I used to watch the clock, counting the hours, even the minutes. The worst part was the stale office air. The building had hermetically sealed windows and a forced air ventilation system that drove me borderline crazy. The perpetually recycled air was like slow asphyxiation.

At break time I would flee the tomb, race down two flights of stairs, leaping three or four at a time, to ground level and out into the sunshine to fill up my lungs with breathable air. The other office workers, meanwhile, eager for their nicotine fix, would light up and stand around blowing smoke at one another as they engaged in idle chit-chat.

One morning during the first week of June, I was on break when I literally ran into Carl in the north lobby. Or, rather, he ran into me. I was coming out of the men's room when he hit the door from the other side with a full head of steam, knocking me six feet backwards. I landed hard on my tail and lay there sprawled on the terrazzo.

Olsen stood in the doorway, a look of surprise on his grizzled old face. He gave a grunt. "Tom Lacey, what are you doing down there?" he shouted. "Speak of the devil."

I felt like giving Carl a piece of my mind but instead, just sat there glowering at him. Olsen stepped over, did his business, zipped up his fly as he moved to the sink, turned on the tap, and began lathering his hands. "You gonn' *make* it?" he said.

I was already on my feet. "What's it been, Carl, two years?"

My own voice sounded like an echo. You see, running into Carl had dredged up some painful memories. Talk of Nolan Leadbetter was out of the question. The man was a ghost. We just stood staring at the floor. Strange what you recall in such a moment, unrelated things, minutiae, vivid despite the passage of time.

Seconds passed. Finally, Olsen sneezed. It came out of nowhere and was so sudden he nearly lost his glasses.

"Gesundheit."

Carl resettled his wire-rims on his nose, then, pulled out a much-abused handkerchief and blew with decision. He wiped his beak one way, then the other. Finished, he stuffed the wad back in his hip pocket. "Memories," he said in an offhand way, as if thankful to be done with them.

"I'm trying to recall the last time I saw you," I said to him. "You don't get into town much, these days, do you?"

By now we had relocated to the hallway.

"Nope," the old man growled in a tone that said, "no further comment." But he added with a wink, "Only this week I'm visiting my daughter and grandkids in Greeley."

"So, how've you been?"

"Not so good."

"You haven't been sick?"

"Last February they opened me up like a fish, here to here," he said, motioning across his abdomen. "Removed half my duodenum. Then, stapled me back together like a Christmas turkey." After a second, he added. "They should have installed a zipper..."

"A zipper?"

"Cancer, you know..."

"Oh Jeez, Carl, I..."

Olsen raised a hand. "Whoa ... Not so fast."

"Huh?"

"Do I look deceased?" The old man chuckled to himself as he tucked in a shirttail.

"Sorry, Carl..."

The smile vanished. Olsen bristled. "Well, smart-ass! You asked me how I was and I told you. That doesn't make me a charity case!" But the dark look passed as swiftly as it had appeared. Carl's face softened. "Heck," he said, "I may not be a spring chicken but most days I feel pretty damn good. Naw, I got nuthin' to complain about."

"You *look* good." It was true. Despite the wear and tear of seventy-odd years and the slow tumor corroding his insides, Olsen looked remarkably well. He had a healthy color and exuded vitality.

"But enough of that," he said. "What *you* been up to?"

I told him about the courthouse job and that I was not happy with it. "I've decided to look for something better," I said.

"They'll snap up a smart young man like you. A fellow with a college degree and all."

Then, I told him. "Carl, I walked away from the degree."

Olsen grunted. "You walked away? Well ... why?" Blunt language is one of Olsen's quirky habits. His flinty style often causes sparks, a style I had admired; but not that morning.

I just glared back. If there is anything I hate it is being grilled about why I failed to pick up the degree. "You knew I changed my major..." I said to him.

"No..."

"Yes. *After...*" I let the words hang, so there could be no mistake about my meaning. "...from zoology to philosophy."

"Hmmm." Another grunt. Silence again drew us apart. Carl turned his head to the side and coughed, then lifted his head slightly. He squinted at me through his bifocals, his eyes narrowing to rivets. "Philosophy, huh? Don't tell me. You ain't one of them *purists?*"

Carl has a way of turning a phrase. Maybe it was his distinctive tone, so unique to his way of speaking, and I mean caustic. Like vinegar. Whatever. Anyway, the absurd moment hit me sideways. I started laughing and could not stop. I remember grabbing Carl's arm to catch my breath, but was I immediately seized again. So it went.

Olsen was not amused. Wary of the rashness of youth, he watched all of this with a stern eye and a disapproving frown. To say he was put out would be an understatement. But there must have been something contagious about my fit, because by degrees the intangible "something", or whatever, slowly wormed its way through his tough old hide. By and by, despite himself, against his own better judgment, Carl warmed to it. When he finally let go and joined in his paunch started heaving like a big bag of Jell-o.

The common cause was just as well, as it saved me from having to tell the rest of the story about my failure to pick up the degree, and even more emphatically, my deep disillusionment with that proud citadel of reason, I mean the university, with its vaunted pathways to higher learning.

Don't mistake what I'm saying. I have no problem with reason or higher learning. But I did and do have issues with the perfunctory professors, the men who preside over the dissemination of knowledge. So called. Brilliant fools, with a few notable exceptions, one being Leadbetter. After four years of classwork I had learned, you see, that there is more than a little truth to the old adage: *Those who can, do. Those who can't, teach.*

In the end it failed to hold me. By the start of my senior year I was in open rebellion and when the time came to graduate I did what a great many college seniors only dream of doing. I followed my stars.

On the day of the commencement I went out and got shit-faced, a wild day and a night of which I remember nothing, from approximately the moment I climbed up on the table at the Red Garter waving a frothing schooner of Coors to recite my epitaph to all of that.

Blotto. The only time I ever woke up in an alley.

The old man caught a feather in his throat and fell out coughing. The geezer seemed prone to it. His eyes looked like they were about to bug out of his head. I clapped him on the back a few times. When he finally came out of it Carl stared at me with something like incredulity.

"So," he said. "Arghhugh-ghmm." He had to clear his throat again before he could continue. "Arghummmm. So let me get this ... umm ... straight. You're telling me that, uh, after four years of college you got no degree, no career, and no, uh, plans or *anything?*"

"That's about the size of it."

He ran a rough hand over his scratchy beard. Carl's jowls jiggled. He seemed to be figuring. Reaching up, he removed his fedora and palmed his thinning scalp. "Jesus, Mary and Joseph," he whispered. Replacing the hat, he rubbed the stubble on his chin and looked me over like he was fitting me out for a new suit or a coffin. "Well dammit, I could use another man, provided he can heft a saw. Are you green?"

"What do you mean?"

"Can you hold up your end of a chainsaw?" he nearly shouted.

"Carl, I've never run a chainsaw."

The old man frowned. He appeared to be thinking. When he spoke next his voice was like thunder. "Well, a man's got to start *some* damn place. All right, all right, tell you what.

I got near eighty acres of pole timber to thin this summer, before the snow flies. Lodgepole pine. It grows like a weed, thick as dog hair or spring grass. I've got a contract for as many corral poles as I can send down the Interstate."

As I listened I remembered something. Yes, about how the old man had come out of retirement. Six months after closing down the professional hunting guide business that had made him a household name in rural Colorado, Carl changed his mind and went back to work. Not as a guide though, his hunting guide days were over. But Carl's the sort who needs to work to be happy. Work is in his blood. Put a man like that out to pasture and you might as well sign his death warrant.

"Oh yes," I said, "I remember hearing ... something about.... Didn't you start up another business? Posts and poles, wasn't it?"

"That's right. I deliver fence poles all over, Denver, Colorado Springs, Omaha, Wichita, you name it, as far east as Kansas City. And I'm looking to hire a coupla' fellows. It's one reason I'm in town. I need cutters. Guys who can hack it. Men who can go a week without a bath and don't mind dust in their oatmeal. Get the picture? Guys who ain't squeamish, or scared of bears in the woods. I'm talking seasonal work you understand, nothing permanent. Probably I can keep you busy into September. I don't pay like the big outfits. I can't. I'm a small-time operator. But it's honest work and I pay in cash. Interested?"

As I recall, I said nothing. The offer had taken me by surprise.

"If you want to...uh, why not drive up next week and give 'er a whack?"

"Up to North Park?"

"Right."

"Well..."

"Tom, it's great country. You oughta' know. You've seen it."

"I know, Carl. But I'm not equipped. I don't own a saw."

Olsen brusquely dismissed the objection with a wave of the arm. "No problem. I got an ol' Homelite you can borry. We'll fix you up, gear wise. Don't you worry. You can work it off. What do you say?"

But I still hesitated. I was thinking.

The old man's demeanor changed with astonishing swiftness. Bristling, he doused me with vinegar. "So, what's the hold up? You said yourself you're looking. You think you're too good for real work? Is that it? Huh? You think you're above a little sweat? Maybe you can't handle an honest eight-hour day? Hmmph! You got fluff between your ears like so many youngsters do these days..." He exploded, "The whole damn nation's in the ditch, gone to the dogs. So which *is* it?"

I never did understand exactly how I made the decision, that morning, whether it was the acid in Olsen's voice or the menacing expression on his weather-beaten old brow. Either could have curdled fresh milk. But the truth is, I *didn't* have a thing to lose and I was genuinely intrigued by the offer. There are worse ways to spend a summer than camping in the mountains.

"OK," I said. "Why not? I can't think of a reason why I shouldn't."

"Good." he said. "Good. Now you're talking, Jim Dandy." Olsen was beaming, clearly satisfied. No more vinegar. Another wink cinched the deal.

TWENTY

Three days later, I loaded my gear into the back of my decrepit Toyota pickup and motored up the Poudre River Highway to Cameron Pass. I knew the road quite well. I had driven the pass before. The eastern approach is steep but has no switchbacks. You down shift the last six miles. The final grade follows a long, slow ascent to the summit. As always, I was stunned at the first sight of the Nokhu Crags looming in the West, jagged, snow-capped, pristine.

The descent clings to a vertiginous cliff for miles before descending to the Michigan River. Five miles west of Gould I took the Rand cutoff, a shortcut to the southern end of North Park; and motored first through lodgepole, then, scrubby sage and occasional aspen stands, until I came to a high point along the shoulder of Owl Mountain. It's a long ridge that intrudes into the southeastern portion of the valley. From the top the view is incredible. The Rockies dominate the horizon in every direction. In the west the snowy Park Range was aflame in the morning sun – the highest peaks suspended above the valley like a glistening crown of jewels.

I descended again and the panorama disappeared from view. The road narrowed to dirt and finally ended at the whistle-stop known as Rand, where I regained the main highway. I headed south and within a few miles started the roller-coaster climb toward Willow Creek Pass.

Twenty minutes later, near the foot of the pass, I turned off the highway onto a Forest Service access road and had no trouble finding Olsen's landing a quarter mile above Snyder Creek.

Olsen was on the deck, apparently working on an antiquated crane which he had partially disassembled. Equipment and tools lay scattered around him. Carl was dressed in a ragged t-shirt, faded jeans and bright red suspenders. On his head was the general purpose fedora he'd probably worn for twenty years. His summer quarters were not much, a tiny trailer parked at the edge of the landing. But with the Never Summer Mountains just two miles southeast of camp he had a spectacular living room.

First thing, Carl sent me packing down along the creek to pick out a suitable campsite and unload my gear. That done, I returned to the trailer. Carl met me at the screen door, a toothpick in his teeth. He had just finished lunch; and promptly handed me a beat-out old Homelite and said "Let's go." Picking his molars, Carl led me out into a patch of sapling-sized lodgepoles. As we walked he explained about keeping "the stumps under six inches" and "the slash under eighteen."

When we were deep in a pole patch Carl reclaimed the saw and started it up with two sharp pulls on the cord. I watched as he dropped a spindly tree, then showed me how to use a logger's tape to measure it to length. A logger's tape resembles a carpenter's tape except that it fastens to your belt. Olsen had customized it by affixing a specially bent horseshoe nail to the end of the tape, for easy attachment. He planted the nail in the butt of the downed tree and began methodically trimming branches with the saw as he moved up the sapling. The tape unreeled as he worked.

He handled the chainsaw with surprising dexterity and economy of motion. I was impressed. When he had done with the trimming Olsen grabbed the tape in his left hand. Leaning down, he stretched it tight against the tree and found his mark. Eyeballing the spot, he let the tape go and cut the pole to the desired length. He flipped the pole over with his boot – a nifty move – and trimmed the backside. *Voila!*

In short order the tree had become a sixteen-foot corral pole. When he was done Carl gave the tape a flick of his wrist to release the nail. The reel was spring-loaded and the tape came whipping back as it rewound with a loud *z-i-i-i-i-n-g* that made me duck.

Olsen found this hilarious. His technique had been honed by years of doing. He repeated it again with a second tree, then handed over the saw and tape, and stepped aside to watch, still picking his teeth, as I dropped my first tree.

I was tentative, at first. But after some coaching the old man decided I was ready for a solo run. At that point, he went off to attend to other matters.

The trial with the saw passed without incident, except that I dipped the blade in the dirt a few times, which rocked several teeth. A half-hour later Olsen returned with a can of gas, bar oil, and a round file in hand, as if in expectation of the worst. He proceeded to show me how to sharpen the chain. The lesson with the file did not deter his young recruit, me, from asking the insistent question.

"OK. But how *often* do you sharpen it?"

"Olsen stared over his glasses with a look so formidable I knew I had asked a dumb one. The patient tone of his voice took some of the sting out of the look. "Whenever it gets dull. *OK?"* (Stress on the "OK?")

"Right."

"Is that too hard to remember..." Olsen's left eyebrow cocked as he unloaded with the whammy, "for a college drop out?" The weather-beaten old face was grimmer than hell, but one eye twinkled insanely.

"No. I *got* it."

"Gooood."

The rest was practice. Nothing fancy, just simple repetition, which is the path to perfection, or, at any rate, mastery. So it went. By the end of the first day I had the basic moves down and was finding the rhythm.

It was not until mid-afternoon of the following day, though, that I discovered I was enjoying myself. At some point everything just...sort of...clicked, as if a switch had been thrown. Suddenly I was in the groove, digging it. Despite the insane racket there was no hint of anxiety or worldly angst, only a perverse feeling of contentment. The honeyed moment came as a monumental surprise because it ran counter to my every expectation. Never in a million years did I suppose I would actually enjoy running a chainsaw.

What surprised me most was how easily it came to me, how quickly the moves became second nature. Stacking the poles was also more enjoyable than I had expected.

I could not help but wonder if I had been born for this.

"Jeez, what a surprise, I told the old man after day two, making no effort now to conceal my enthusiasm. You know, I thought this type of work was, uh, I don't know what I thought. Grunt work. Drudgery, I suppose. But, hey, it isn't half-bad. Olsen was grinning from ear to ear. To tell you the truth, it's kind of..."

He cut in. "Like a happy kick in the shorts. Right?"

"Right. But how did *you* know?"

Ripping off his hat, the old man bent over and roared with laughter; which set the tub of lard in his belly to quivering again. "You have to ask!"

Living out of a tent in the woods and cooking over an open fire was not half bad, either. To tell you the truth, the practical simplicity of it all was much to my liking.

There was, of course, a learning curve. Success with a chainsaw, even survival, depends on adhering to a few basic rules, which are mostly common sense. Fortunately, I was a quick study.

Even so, I had a minor setback the morning of the third day when I cut my knee. It happened while trimming branches. I accidentally moved my leg in too close to my saw's twenty-two inch bar. Somehow my jeans, the cotton fabric, became entangled in the chain, which instantly pulled the spinning teeth down into my leg. Thankfully, the cut was superficial, nothing serious. The gash was shallow, no more than a flesh wound. But it did require several stitches.

When I showed Olsen the damage he disappeared into his trailer and reappeared with a sewing kit and a bottle of peroxide. It was, he said, a case of on-the-job training. "Have fun," he said, smiling wickedly as he handed it over.

"You mean, you're not going to sew me up?"

"Do I look like a nurse?" More battery acid.

I threaded the needle, applied the peroxide, and though the parted flaps of skin made me shudder, nonetheless, I stoically stitched them together, gritting my teeth with each stitch, grit in place of anesthesia. What was the alternative? The nearest medical clinic was in Laramie, three hours away – too far to drive for just a few stitches.

The next morning I was back to work, a little stiff in the joint, but none the worse for it. If anything, the mishap was a blessing, as it taught me an important lesson. I had learned the serious consequences of getting too near those razor-sharp teeth.

The slip with the chain never recurred.

There were no other surprises. I quickly learned to eyeball the proper spacing, though it took me several more days to master the art of selecting the best 'leave tree,' in other words, the lodgepole with the fullest crown. By this time I was already picking up speed. By week's end I was moving through Olsen's patch of dog hair like a one-man-swarm of pine beetles.

I recall that, once, I happened to look up in the thick of it and noticed Carl standing off to one side, watching me work. The old man was grinning like the Cheshire cat. I knew then I was home free.

The following week another hired man showed up, a seasoned cutter named Dave Roper. He is a giant of a man, about six-feet-seven. Roper had missed the start-up because he had just gotten married in San Francisco. He and his young bride, Nancy, were still on their honeymoon. They established their own camp in a lodgepole stand about a mile above the landing. Roper was a woodsman and, like me, felt right at home in the mountains. But Nancy was from Omaha, a city girl and hated "roughing it."

Within the week the honeymoon was over.

I shared my evening meal with them several times, until I became embroiled in one of their bitter quarrels. Dave and Nancy's arguments were soon the regular horror.

It did not take Roper and I long to cut our way through eighty-odd acres of lodgepole pine. In late August, when the final sloping piece of ground was done and the last poles had been stacked on the landing, I decided I was ready for the big time. With part of my pay I bought a new pair of boots (caulks or, as they say, 'Corks') and a spanking new Swedish saw with a larger bore. Then, I jumped over the divide and hired on with Jacques St. Clair who had a big logging contract underway about thirty miles southwest of Granby. That was how I moved out of the poles and into the big timber.

Tallie had been listening quietly all this time. "But what about your adviser? You said he was a ghost. What happened to him?"

In late spring of my sophomore year, Leadbetter asked me to accompany him on the final ground-truthing phase of his prairie dog study. I was honored to be a part of the expedition; and it was indeed memorable, probably the high point of my university education.

Outfitted and led by Olsen, the three of us spent that entire summer in the field. We made extended forays into Colorado's high valleys and wilds that are the last bastions of the prairie dog. We started in the San Luis Valley, then moved around the state. This included sweeps through South Park and North Park during which we gathered many samples and catalogued a dozen previously unknown prairie dog towns. In August we moved west into the White River country. The White is one of the main tributaries of the Colorado.

The summer passed in what you could call idyllic splendor. In mid-September we completed the last of the fieldwork during a two-week trip, more like a working vacation, through southwestern Colorado. Most of the time we were within sight of the San Juan Mountains. There was plenty of time to catch some trophy-size trout.

The summer ended on that high note, after which we moved indoors. The project continued on campus in the zoology lab located in the Agriculture Building. There were fresh specimens to be mounted, a thousand photographs to be sorted and collated, a mountain of raw data to be analyzed.

Most days I assisted a graduate student named Jeff Sherwood, brought in from the physics department to do the statistical work and develop a mathematical model.

But even before the number crunching was done, we knew the results would bear out Leadbetter's preliminary assessment of the status of the prairie dog in the state. No mistake, all three varieties were in a steep decline throughout their range after a century of human depredations. Ranchers and farmers regard prairie dogs as a nuisance. The use of poison is widespread and has been for many years. Urgent steps were needed to protect the last strongholds.

How well I recall those last days. Leadbetter was at his ebullient best. The journal *Nature* had commissioned a feature article about the threatened prairie dog for its upcoming January issue, set to run with a color photo centerfold. The publisher had sent word that a close-up shot of a prairie dog would even grace the cover. Several other scientific papers were also in various stages of completion, slated for various zoological journals.

I had never seen my adviser in higher spirits as he fussed over loose ends and busied himself with the write ups, that is, when he wasn't delivering one of his brilliant lectures on vertebrate paleontology. Leadbetter had a following on campus. His classes were usually attended by standing-room-only crowds.

On the 14th of November, I will never forget, I dined with Leadbetter and his attractive wife, Henrietta, who treated me like family. After dinner, over coffee, the professor and I discussed my Master's thesis. I was still an undergraduate, but he had often encouraged me to plan ahead. Leadbetter's enthusiasm for science was contagious, and that evening he wanted to know if I was ready to submit a formal proposal to the faculty committee, which was scheduled to meet the very next day. I was thrilled. Imagine my surprise. It was one of the most exciting moments of my life.

However, when I arrived in the morning at eight sharp, briefcase in hand, I found Leadbetter's office cordoned off with yellow ribbons, as if the place was a crime scene. Graduate students and faculty were milling about in the hallway. They looked dazed, as if in shock.

I waited until the cop by the door turned his back, then ducked into the inner sanctum. I was not prepared for what I found.

Earlier that morning a custodian had discovered Leadbetter in a pool of his own blood. On the desk was a suicide note that explained nothing. On the floor – an empty revolver. He had fired one bullet.

The question that arose in my mind at that time still haunts me, for I have never answered it. Why would a man at the summit of his career, a man with everything to live for, put a .38 in his mouth and splatter his brains across the green pastel wall of his office? Why?

It was the beginning of the end of my academic studies. Something had fundamentally changed and there was no going back. The loss of my mentor, the professor whom I most respected, plunged me into a life crisis unlike anything I'd ever known. A crisis of faith.

The day of the funeral, I flung my thesis proposal into the trashcan. I was done with all of that. Zoology now seemed so pointless. The next two years was a time of darkness. At times, I felt I was losing it. I changed my major to philosophy and began the search that still preoccupies me.

Next to the grim finality of death, life seemed so fragile and illusory. How could anything so tenuous have real significance? How could anything so fleeting have meaning? And if life has neither significance nor meaning, what then? Does nothing matter?"

TWENTY ONE

"Your letters never reached me," she told him. "Florence and I have not spoken since the day I arrived in San Francisco, and showed up at her flat. We don't talk anymore." She described their heated row, after her erstwhile friend, strapped for cash, had sub-let the spare room that she supposedly had been saving for her. As a result of which, Tallie found herself on the street in a strange city with no lodging, almost no money, and no friends – a scary prospect. It was her first visit to the West coast and she did not even have a map of the city. The stress triggered one of her worst migraines ever. In her agony she did not know what to do and, almost blind with pain, lay down on a crowded sidewalk. It was rush hour, and people passed her by or stepped over her, until a woman named Jane came to her rescue. The woman who happened to be a nurse helped her up, and they limped to Jane's nearby apartment where Tallie remained until the migraine ran its course. Somehow this kindness from a complete stranger was transformative. With Jane's help, Tallie found a job in a local bookstore and a room near Golden Gate Park, and started a new life.

She soon learned to love San Francisco. By then, it was February. Spring was already underway. She was pleased to discover that the city was one big public garden; and soon developed the habit of going for long walks. She liked to climb the hills for the spectacular views of the city and the Bay. She loved the bracing ocean air and the pungent sea wrack; she even enjoyed the fog. Her neighborhood was bustling with young people. She met several artists and street musicians and began to go out and have fun. The city was loaded with great restaurants. There was live Jazz or Rock nearly every night. Take your pick. And she spent many hours exploring the city's fine museums. But her favorite pastime was riding BART and the cable cars to the end of the line. Alone.

"No boyfriend?"

"Nu-huh. I needed to....figure things out."

But she never answered his question, "Why didn't you write sooner?"

Later that afternoon she took him out bareback riding on old Luther, a cream-colored Arabian stallion. Luther was one of Mary's oldest stud horses, and her favorite. The twenty-three-year old stallion was half blind, but otherwise in excellent condition. Tallie led him by a rope hackamore as they went out the long drive. Tom opened the gate. They crossed the lane, passed through a second gate and headed for "the south forty." Whereupon, Tallie grabbed a handful of mane and slipped up behind Luther's withers like she'd been weaned in the saddle.

He was amazed, all the more so because of her slight frame. As he watched her ride old Luther he recalled what Mary had said, the previous evening. "Tallie? Ah, even as a child my niece had a natural seat. Spent her summers at the ranch, and most of her time around the horses. Rode before she could walk, usually bareback. Hated saddles as much as shoes and hardly ever used one. A born rider that niece of mine."

As Tallie took Luther around the meadow at an easy canter he realized she was wholly in her element.

After the warm-up she nudged the horse and gave him rein. Luther needed little encouragement and flew around the meadow kicking up clods. After two more laps she brought him up hard, reining him in at the last moment. Tallie slipped down effortlessly and handed him the rope.

"OK, partner, your turn. Remember, it's balance, not grip. He's ornery. If he lowers his head you pull up with the halter so he can't buck. OK?"

"Right."

"Tom, what did I say?"

"Balance you said, not grip. Pull him up."

"I'm telling you, he can be ornery. Watch his head. Use the halter. Pull it up fast and hard if he gives you any trouble."

"No problem," Tom said.

She formed a stirrup with her clasped hands to assist him. Luther was lathered up and game for more. Tom placed his left foot securely in her hands, then, grabbed a handful of mane. The other leg went up and over. But before he knew what was happening the hard ground came up and whacked him. He rolled over staring up at blue sky. The horse was nowhere to be seen.

"Whaaa?" Wuzzies danced around his head. It was only then he realized that the stallion had thrown him.

The blind old horse had shooed him off as if he were no more than a troublesome fly. Tom lifted his head into swirling mist and let it down again, easy-like. He pulled some grass out of his mouth, and coughed. Closing his eyes, he took a deep breath and lay very still. Things were still spinning. Finally, he sat up. The fog parted. Luther was grazing contentedly no more than fifteen feet away, the rope hackamore dragging loose on the ground. Now, came the biggest surprise. Tallie was standing over him smiling ear to ear, on the verge of busting out laughing. Her freckles had gone that silly shade of green, again. He was not amused. He snorted a sprig out of his nose. Light began to penetrate the clouds. She evidently felt a twinge of guilt, because suddenly she was cooing, all bothered and concerned.

"Oh! You poor thing," she fretted in a soothing voice. She knelt down, brushing back loose hair from her face, and said earnestly, "Are you sure you're OK? Geeze, Tom, I'm sorry. I should have warned you. Here."

She helped him to his feet and steadied him as he stumbled about. His legs were rubbery. But no harm had been done. Nothing was broken. He was all in one piece, except for his ego.

To her credit she did manage to keep a straight face, but only just. It was her brown eyes gave her away. She had pushed her sunglasses up over her forehead. She could scarcely keep it in. "I just wanted to see if you could handle him." Her attempt at a pout was pretty, but hard to take. That was when he noticed the green freckles going up and over the bridge of her cute little curled nose.

He found out later that due to an abusive incident, many years before, old Luther harbored a deep grudge against men. The stallion had not been successfully ridden nor even mounted by a man in more than ten years. She might be slender, lost in a dress, just a slip of a thing, but Tallie was no less the pure bred article. Any woman with buck like that would make a man proud one day, *if* he could manage to stay in the saddle.

"Tallie," he said. But he never got the rest of it out.

"I want to show you something."

"Show me? What?"

"One of my favorite places."

He mumbled something incoherent. "I'm not sure if... uh." She had mounted Luther and now extended her hand.

"Oh come on..."

"Not a chance. I'll walk. You ride."

"See, it's perfectly safe. As long as I'm with you he won't buck. Promise. Here. Give me your hand..."

"Well ... I don't know." He hesitated for a moment. "Oh, what the hell." He took her hand and swung up. Luther appeared not even to notice.

Up they rode along a horse trail that followed an arroyo to the head of a wash. They dismounted in the shade of junipers for a cool drink at a clear water spring. Above the spring the trail grew steeper and they continued on foot. Tallie led the horse and Tom followed, through a forest of rough-barked Ponderosa pines that eventually opened into a small meadow. Sunlight was filtering into the glade through the surrounding trees. Now, they climbed through wild flowers and high grass. The air was filled with butterflies. The trail grew steeper. They walked through fragrant horsemint and wild iris.

Presently, a small shack came in view.

"Trail's end," she said. It was not a cabin proper, just a rough hut with tarpaper siding set on wooden piers. The place had a decent roof, though. Outside there was a block for splitting wood. She hobbled Luther and they went in.

It was a single room with the barest rudiments. For heat a small potbelly stove set on flagstones in the corner. An axe was propped against the wall. A pile of split wood lay neatly stacked behind the door. The double bed took up most of the room. Actually, it was just a faded old mattress on a makeshift frame at the level of the large picture window. The mattress was water-stained and leaked stuffing at one corner. The room had a musty smell. Along the north wall was a small table with two chairs that did not match. On the table was a vase-full of last year's faded flowers, an ashtray with butts and roaches, and a small urn, the kind used for incense. Beside it lay a fragile line of ash. On the wall was a black and white photo of Janis Joplin, postmortem queen of the counter-culture. Janis had her hands on her knees and was smiling in full hippie regalia, strings of beads dangling from her neck. Tacked to the ceiling over the bed was a sagging poster of Jimi Hendrix, guitar in hand, frozen forever in mid-chord. On the dusty sill were stubs and partially burned candles of various lengths, one with a parted lip and a thick wax puddle at the base. A multi-sided crystal of cut glass hung by a thread, with a propensity to spin and cast its prismatic rainbow about the room.

The window commanded a terrific view of Trail Ridge fifteen miles to the northeast. A lazy fly buzzed against the windowpane. The only other noise was the sound of their breathing. They were winded from the climb. The look of Janis the erstwhile hippie queen said *deja vu* all over again.

Now she was beside him. Gently she brushed her ponytail back over her shoulder.

"Thanks," he whispered.

"For what?"

"For coming."

Her small hand found its way into his. When she kissed him his heart fluttered, before leaping out of his chest.

TWENTY TWO

Bright and early Monday morning a delegation from the Ancient Forest Rescue paid an unannounced visit to the logging camp in the Kawuneechee Valley. The green tide was a half-dozen strong and was led by Pinecone Peters. The visit was born of a new sense of urgency. The campaign to save Bowen Gulch was in danger of being overtaken by events. Time was running out. The delegation hoped to gain a reprieve. If even a few of the loggers could be persuaded to desist or delay...

Pinecone had learned about the camp through dumb luck, from one of the loggers, no less.

The ad hoc plan was to politely introduce themselves, then engage the residents in a rational discussion about ancient forests and especially the ecological need to preserve them. The arguments were cogent. The facts were science-based and were overwhelmingly on the side of preservation. With truth on their side, how could they fail?

Pinecone and his cohorts began knocking on camper doors and windows, going from one to another.

"Hello, there? Anyone home?"

But it was a ghost camp. The green team had no way of knowing that Jacques' loggers were not due back from the long weekend until the following day. Far from being deterred, Pinecone and company did the next best thing. They began to disperse their literature freely, slipping activist fliers under doors and brochures into open windows. One young woman boldly peeked under the canvas flap of a wall tent. Finding it deserted, she left a neatly folded handbill on the cot, a personal touch. Pinecone was busily scotch-taping a flier to a camper door when Wolfe Withers suddenly appeared from around the back. Wolfe had his morning coffee in hand and was chewing the last of his buttered toast.

"What can I do you for?" he said in a deadpan voice, with no hint of emotion. Wolfe had on his perpetual scowl. His face conveyed a challenge, unambiguous and emphatic. A wintry gust passed over the activists standing behind Pinecone who nervously looked at one another as if searching for consensus about how to proceed.

Pinecone stepped forward. "Good morning, sir," he said with a pleasant smile. He handed Wolfe one of the handbills. "We are with the Ancient Forest Rescue and we ... uh ... were hoping to chat. That is, if you don't mind." Wolfe gave no reply, only a cold stare. After a pause Pinecone continued. "We uh, understand that you are ... uh ... scheduled to start cutting in a few days. Up at Bowen Gulch, and we..."

Wolfe had taken the flier without a downward glance, his eyes boring straight ahead through Pinecone like a laser. Now, he one-handedly crumpled the flier into a wad, then, with a quick upward jerk of his wrist flung it back in Pinecone's face. The wad had a low trajectory and bounced off the bridge of his nose. Bingo. The delegation winced. Pinecone was as startled as the rest. Caught off guard.

A gauntlet had been thrown.

It took Pinecone a second or two to get a grip, but in the end he handled his anger, and extended his hand as if to calm down the fast-developing situation. "Look, sir," he said, "we don't want any trouble..."

The words "sir" and "trouble" or rather, the thought-form of the words, electrified the air like a proleptic archetype.

Wolfe was in the habit of venting for fun and pleasure, usually on human punching bags. Although not a boozer he had been known to hit the bars just to stir up a fight. No special occasion was necessary. Any old excuse was cause enough, especially this unwelcome visit by these uninvited tree-huggers.

Wolfe took a step forward, swelled out his massive chest, and got on with it. "What say motherfuckers?" The big man had set his feet and now beckoned with his free hand.

Nary a sound. Stares of disbelief. Wolfe loudly cleared his throat, spat on the ground, the better to address them, and got on with it. "Well, don't all of you little wussies cry at once."

Silence.

"What's the matter, groupies? Cat got your tongue?"

More silence.

"Heh-heh-heh. Well oh my, oh my. What a sight. Suits me Jim Dandy. You are one sorry bunch."

Pinecone managed to say, "We did not come looking for a fight."

Wolfe chuckled in his scornful way. "Well sheeeeit. I'll live to piss on your freaking graves." To make the point he dumped the dregs of his coffee, fouling one of his boots in the process. He stroked his chin and scoffed his last. "Hell, I'll do it now."

Unzipping his fly, he pulled out his pecker and proceeded to empty his bladder on the ground. "Mmmm that was one fine cup of coffee..."

The measure of his disdain rose and arced quite gracefully in the direction of Pinecone and the rest, who by now were beating a hasty retreat to their vehicles, occasionally glancing back over their shoulders.

"At's right! Run home to momma!"

Wolfe chortled as they dispersed. He would have preferred to get physical but it could wait. No hurry. Another time, another day, he would get his chance to stomp some hippie ass.

TWENTY THREE

Jacques St. Clair was down on one knee beside his dozer when the Forest Service truck appeared on the landing. He glanced at his watch, then rose and stood with his hands on his hips as the driver climbed out of the olive green pickup and strode toward him.

The chief accepted the fact that federal employees who administer national forest lands should drive green pickups. Green was appropriate. But he always wondered about the lame shade no doubt chosen by some bureaucrat that resembled green puke more than any known conifer. St. Clair had dealt with the U.S. Forest Service on countless occasions over the years, and had long since come to view the sickly green as apropos of the agency as a whole. It was a damning observation, but on par for a bureaucracy that could never decide what business it was in.

Jacques was of the opinion that no good ever came from trying to please everyone. It was bone-headed even to try, because such an attitude was invariably doomed to fail. Yet, this summed up the government's official policy of "multiple use." Jacques found it difficult to conceal his contempt for men who tried to be all things to all people, which in his view betrayed a fundamental weakness of character. Embody one man's foibles in a government bureaucracy and you had a disaster in the making. No wonder the nation was going down the drain.

Jacques was rankled by the fact that his right-of-way business often compelled him to work closely with this agency that he so despised. However, in life one had to go along to get along, and that meant stowing one's personal opinions.

He took some consolation in the fact that most of the agency staffers were friendly and easy enough to deal with, including this fellow walking toward him, Bill Noonan, with whom he was well acquainted. St. Clair got along just fine with Noonan, because the man was a timber beast.

"Morning, Bill."

"Jacques," the other man said. He nodded toward the tractor. "Problems?"

"Yeah," Jacques said as he adjusted his cap. "But nothing we can't handle. Busted track. Hang on." He crouched down to confer with Francis, whose feet and legs were sticking out from under the cat. Assorted tools were scattered about him, including wrenches, a cutting torch, a sledgehammer, a cable-handled punch, and a heavy pry bar made from an old truck axle.

"You see the master pin?"

"No such luck, boss. We're gonna' have to cut some pads loose."

"Dammit. I was hoping we could avoid that. Oh well, fuck it. Go ahead and do it."

Space was cramped under the cat but Francis nonetheless moved easily into position. With the oxyacetylene torch in one hand, he opened the gas valve with the other. Then, he sparked a flint-striker. On the third try the tip ignited, producing a lazy yellow flame. Letting go of the striker, Francis twisted another valve, which fed oxygen to the flaming tip. Instantly the sloppy flame became a hissing torch. One more adjustment narrowed the flame into a brilliant blue focus. He was ready to cut steel. Francis pulled his dark goggles down over his eyes and applied the torch to one of the heavy dozer tracks. He held the tip steady along the outer edge until the steel was cherry red. Then, he hit the oxygen and the plate gave way amidst a shower of brilliant sparks. The man showed a soft touch as he guided the torch across the heavy plate. The steel parted like hot butter. As he worked Francis stared directly into the heart of the intense blue flame.

St. Clair and Bill Noonan instinctively turned away from the blue brightness to save their eyes. Noonan slapped some rolled papers against his leg.

"I see you brought the maps."

"There have been a few changes. Care to have a look?"

The boss motioned with his head. "Yeah, sure. Step into my little chamber of horrors."

Noonan followed Jacques into the office trailer where he cleared the wide desk so that Noonan could spread out the maps. Twisting the top one around, he planted his index finger. "This is your landing. We are...here." Noonan then explained how the sale was laid out, where the various units were located, where the skid roads would be, and so on. The boss had already studied an older version of the same map. The only difference with this newest draft was that one of the original sale units had been dropped, and another added. There had also been some minor changes to the skid roads.

"Is this set of maps for me?" Jacques said, fidgeting with his cap.

"Yes. I will leave them."

"How about some coffee? I could make a fresh pot."

"Sounds reasonable."

"Make yourself comfortable."

There were no chairs and Noonan chuckled at the irony. "Sure. Don't mind if I do." The trailer was not designed for comfort. One had to improvise. Noonan sat on some crates piled along one wall.

While the coffee perked the men discussed the Bowen Gulch sale. The designated prescriptions were not clearcuts, but shelterwoods. This meant that the large trees would be removed in stages. In the first phase about 30-40% of the forest would be cut, with the rest to be logged in two subsequent entries. The sale also included several smaller units slated for commercial thinning, or thinning from below, as it was referred to.

"The marking crew finished up last week," Noonan said.

"Great. I noticed your guys left behind two cases of marking paint on the landing."

"Spray paint?"

"Yes."

"Leave 'em, for now. We'll pick them up, later."

"OK. How do you take it?" Jacques said as he poured out the fresh coffee. "We got black and we got black."

"In that case, I'll take mine with," Noonan quipped.

Jacques passed him the cup. "Anything else I ought to know?"

"Thanks. No...uh, well, yes, there is, actually. We had another demo at the district office, yesterday morning. You hear about it?"

"Tree huggers?"

"Yep. About twenty protesters. You might see them up here too before very long."

"Not a bit worried. We can handle it."

The men were still sipping their coffee when the trailer door opened. "Done, chief," Francis said. "She'll be fine."

"That's what I like to hear."

Francis carried in some tools and put them away in the adjoining room, actually more of a large closet, and returned with a second load, which he also stowed. Then, he stood wiping his hands with a rag.

"Want some coffee?"

"No. I'm good."

"OK. Take the rest of the day off. You earned it."

"Thanks, boss."

"Francis, now I think of it, drop by the camp on your way back to town. Make sure the men know we start tomorrow at sunup. OK?"

"Sure, boss."

When Delacour had gone Noonan slapped his knee. "Jacques," he said, "I see you have your boots on. You up for a little hike?"

Jacques laughed. "Ready? I was born ready." He laughed because from past experience he knew that when Bill Noonan said, "Let's take a little hike" he was playing with you. He knew they would be lucky to return before sundown.

"I'm ready when you are," he said. Jacques grabbed his cap.

"If you have a pair of binoculars," Noonan said, "bring them."

St. Clair slung a pair around his neck, then rolled up the maps and followed Noonan out the door. The two crossed the deck, strode past St. Clair's yellow skidders, and headed into the timber. They followed the survey stakes and flagging which marked the designated route for the main skid road, which had not yet been constructed. It existed only on paper, for which reason they had to bushwhack through heavy timber. The planned skid road would follow the sloping topography, more or less.

When they were deep in the timber, Noonan paused to study the map. He pointed to some yellow flagging off to their left. "See, that is the boundary of unit #3," he said, pointing. "Unit #4 is just ahead. I suggest you start with these lower most accessible units. You'll have to stay on the designated skid trails."

"Looks like you've marked about a third of the big trees."

"Yes, about a third. You will cut only the marked trees. Leave the rest."

"Right."

Jacques was liking everything he saw. The timber was impressive. He had logged large trees before on other projects, but nothing like this. The day before, he had walked a small portion of the sale area, and had been amazed. Now, he was astonished all over again by the girth of the trees. Some of the trunks looked to be more than four feet in diameter, breast height. Around the base they swelled out even more. "Man oh man, there's some big old daddies in here," he said. "And I don't think I've ever seen so many broken tops."

"Yes, most of the stand is over mature," said Noonan. "Some of the spruces in this unit are 600 years old – maybe older. We cored a few of them. Should have been harvested, decades ago."

They transected the slope, slowly gaining elevation. It was rough going. There were occasional patches of snow. Every so often, Noonan paused to point out yellow markers designating the boundary of one or another sale unit. Some of the units were hundreds of acres in size. Jacques asked about several markers that had been 'X'ed out with black spray paint. "That unit we dropped."

The snow became deeper as they slowly climbed higher, and generally west across the southern perimeter of the valley. Finally they reached the top end of the sale area. "This is unit six," Noonan said as he pointed to a line of flagging. "We are standing at the bottom edge of the last and highest unit." The snow was even deeper beyond the flagging.

"We won't get in here before Labor Day."

Noonan nodded. The two discussed the timber sale awhile, then started back. They tramped for about ten minutes down the gulch. Then, Noonan said, "OK. How about a little workout? I trust you're in shape." Abandoning the gentle grade and the ribbons, he plunged into the undergrowth, down a much steeper slope toward Bowen Creek. Jacques followed. Minutes later, they reached the creek, crossed over on a downed tree, heading for the north ridge. The snow had drifted deep along the tow slope where Noonan paused. "From up top," he said, motioning with his arm, "you can see the whole magilla." They were a bit winded and rested briefly. Jacques knew that higher up on this south-facing slope there would be little or no snow.

"OK. Let's do it." Noonan turned and started straight up the side of the mountain. Jacques fell in behind.

From the first the ascent was steep and rocky. They needed handholds, rocks, branches and stems, to gain ground. Occasionally they slipped back. Two steps forward, one step back. They had to stop and catch their breath about every fifty feet of elevation. Presently, the spruces thinned out and they climbed through a pure stand of Subalpine Fir. As they went higher the trees became smaller and sparser. Fifteen minutes later, after tramping over a narrow cornice of snow along the nape of the ridge, they gained the top. They were well above timberline. Much of Colorado lay at their feet.

"Best part of my job," Noonan crowed.

"No kidding. What a view!"

The vista was spectacular, nothing short of breathtaking. Bowen Peak loomed to the west. The rest of the skyline was dominated by the Never Summer massif to the north, the Mummy Range to the northeast, and part of the Front Range due east and southeast. Their vantage was completely exposed. A stiff wind buffeted their faces. Neither man spoke. Talk seemed superfluous. The foreground view-shed included most of Bowen Gulch and, just beyond it, a good part of the Colorado River valley. The men scanned the gulch with their binoculars. Noonan pointed out the approximate locations of the various sale units, and the planned main skid road.

"From here you can see most of the units. Hey, look," he said, "you can just make out your skidders. There's the trailer, too. See."

"Where?"

"Look over yonder, beyond that rock outcrop."

The boss searched with his binoculars. "Yeah, I got it. That's them all right."

"Also looks like some of your crew."

Jacques adjusted the focus. "What? Wait. That's not my crew," he said. "Who are those guys?"

"Did you see that?" Noonan said. "I think I just saw a guy cut a hydraulic line."

"Merde!" St. Clair shouted through his binocs. "I'm being sabotaged!"

TWENTY FOUR

"Broad fucking daylight!" Jacques spat through his teeth, muttering to no one in particular as he paced to and fro beside his dozer.

Tom had never seen the boss like this. Jacques' face was flushed and he had a wild look in his eye as if he were capable of anything.

When word of the sabotage had reached the logging camp the crew hurried to Bowen Gulch, too late. Now they stood staring at the slashed lines and the large pool of hydraulic fluid under each of the skidders. The fluid had spilled everywhere and continued to drip from the cut lines, which hung limp.

The damage to the skidder engines looked to be total. Someone had removed the oil caps and dumped sand and gravel into the crank-cases, filling them to the top. The engines would have to be overhauled at huge expense before the skidders would run, again. The boss estimated his losses in excess of $12,000 and still climbing.

Jacques had been fortunate in one respect though, because bad as the vandalism was, it might have been worse. The detachable metal covers that Francis had installed on both sides of the D-6 engine housing had foiled the saboteurs, preventing them from decommissioning his cat. The evil-doers had been unable to remove the steel jackets or cut the padlocks that held them in place. The protective hood had saved the cat's motor.

"Good thing for the metal jackets," said Francis.

The boss stirred from his gloomy thoughts long enough to mumble, "Mmm-hmm, yeah. Right. Thank heaven for small favors."

Jacques knew he had also been lucky another way. The tree huggers had entirely overlooked the dozer's exhaust pipe, which had been vulnerable to sabotage. They might easily have put the dozer out of commission simply by dumping stones and dirt down the pipe. The stack was wide open, totally exposed. It was a small thing but it brought some relief.

By degrees the chief calmed down, began to think rationally, and took stock of the situation. The sabotage had tossed a serious monkey wrench into his operational plans. There was no way, now, that he could start up the project on schedule. He was facing a long drive to Denver that very afternoon to locate and rent two replacement skidders.

Or try to...

If all went well, the trip would cost him two days. The out-of-action skidders would also need to be loaded up and moved to a garage. At the moment, however, that was not a priority. That unpleasant chore could wait until later; much later, if need be.

Nor was this all. Unidentified men had been seen lurking in one of the lower sale units, precisely where Jacques intended to start the cut. Inspection by Noonan had confirmed the worst. The eco-nuts had been in there, all right, pulling up survey stakes, removing the yellow flagging, and spiking trees. A substantial part of the skid road would now have to be resurveyed. The tree spikes were not only hard to spot, they were extremely difficult to remove.

A spike could dull a saw chain, or worse, cause a thousand dollars damage to a large mill saw in a split-second. The intent of the tree-huggers was straightforward: drive up the cost of logging and reduce the profit and the incentive to proceed with the sale. The spikes also posed undeniable safety hazards to mill workers, though the actual extent of the hazard remained controversial. The dangers were unspecified, and disputed. Horrifying stories of flying spikes had appeared in the papers. In one case, a mill hand reported a metal spike shooting straight up through the roof. It did not take much imagination to scare up the grisly outcome if an unlucky mill worker happened to be in the path of such a projectile. Fortunately, no one, to date, had been killed or seriously injured by a tree spike; but this did not stop the industry and the newspapers from portraying radical environmentalists as cunning terrorists on a par with Abu Nidal.

The practical consequence of the spiked trees was further delay.

The staffers at the Forest Service district office were as angry as hornets about the sabotage, and had debated whether or not to remove the spikes. In the end they settled on a quicker alternative. They would simply drop the unit with the spiked trees and add a replacement unit to the sale. The resurvey job and reflagging would probably take about a week. However, there was no way to know for sure. Past experience had taught Jacques to expect the worst from an agency that usually moved in slow motion. Bill Noonan had promised to light a fire at the district office, but Jacques would believe it when he saw it happen.

The boss was a blur, now, as he barked orders aimed at cutting his losses. First thing, he sent Francis into Granby to make arrangements for his usual low-bed operator to move the D-6 over to the logging camp. Best park it where the crew could keep an eye on it, at least until he returned from Denver. No way was he going to leave his prize cat at the logging site where it would be vulnerable to further depredations while he was gone. Jacques had a feeling the bastards would try again and felt totally exposed. He had called the Sheriff's department, first thing, but the county cops had failed to respond with the alacrity the situation demanded.

That was more than two hours ago. So, where is my protection?

It was a touchy item, near the top of the boss's shit list.

Later that afternoon, the Western-Pacific attorney showed up on the landing without notice in a sleek black limousine, just as Jacques was about to depart for Denver.

It was a late model Cadillac, with darkened windows. Most of the crew, including Tom, was still on hand, and they watched the attorney climb out of the back seat. He looked like the stereotypic high roller, complete with the trench coat and dark shades, even a cigar. The man was obviously none too pleased. He had a grim look on his mug of a face. It seemed to Tom the guy had stepped straight out of a Hollywood "B" gangster film.

After exchanging a few curt words with the boss, the two men disappeared into the office. When they emerged a half-hour later St. Clair's face was pale. By then, Tom had returned to camp. But he heard the full story about what had transpired from Francis Delacour, who got it from the horse's mouth.

Western-Pacific had upped the ante. Apparently, the front office had been alarmed by the eco-sabotage and on short notice dispatched their top attorney. The man had flown in from Denver to take charge of the situation. The attorney spelled it out in blunt language. He wanted the boss to send his crew into all of the marked units at first light and start dropping trees. "As fast as possible. And keep dropping them. We are fully within our rights. We have the contract, we own the timber, and possession is nine-tenths of the law. For now," he told Jacques, "don't bother to push the skid road in. You just get those trees on the ground. You can skid 'em out, later..."

But this direct interference did not set well with the boss. Jacques told the lawyer to his face where to stuff it. According to Francis, "Where the sun don't shine."

That was when the lawyer hit him with the whammy. "Well, St. Clair, let me put it in plain English. If you can't get it done, we'll tear up your silly sub-contract and find another operator who knows how to follow orders. You get the picture?" Then, he twisted the knife, adding the implicit slur, "*Comprende vous?*"

TWENTY FIVE

Along the rear wall of the Nugget saloon was a row of half-lit booths, and in the far left corner an even more dimly lit wrap-around stall with a circular table. The booth was well away from the saloon's overhead lighting, and out of earshot of the usual crowd at the bar. This confidential corner was where Pinecone met regularly with his cohorts, Mike Phillips and Steve Gaylord. Both of his mates were fellow refugees from mainstream environmentalism and shared his sympathies with the more radical Earth First movement.

Each had been actively involved with the Ancient Forest Rescue campaign, a diverse coalition; but they also made up their own separate splinter group. They were a generic enterprise, a splinter with no formal identity, just another nameless cell of a broader movement. After all, a group with a formal name, a paid staff, a phone and a front office becomes a recognizable blip on someone's radar screen; hence, a target for psy-ops. Best avoid that with the camouflage of anonymity. The bard put it best: "What's in a name?"

The trio had no acknowledged leader for similar reasons. Like the Musketeers of old, they were true libertarians and freely passed the mantle of "leader" around the table among themselves, taking turns, usually with a good deal of associated fun and banter. Occasionally they drew straws, but always they operated on a consensual basis. A group with a designated leader, after all, can be decapitated, and so reduced to flailing arms and legs. But a headless (not to say mindless) trio can strike dread into the heart of the enemy, no less than a disembodied rider on a moonlit night.

There was important business on tap, this particular evening. The fight to save the ancient trees of Bowen

Gulch was fast approaching a climax. The contract for the timber sale had been awarded and a dozen derivative deals inked. That very morning, they had received worrisome news. A logging crew was now ready to go. The campaign to save one of the last great places was in danger of being outflanked by modern-day minotaurs, half-human and half-machine, strange hybrid monsters with the cerebrum of a man and jaws of ravening steel. Soon, the fabulous creatures would be unleashed on one of Nature's last unspoiled strongholds. The liquidation of Bowen Gulch's 800-year-old trees was about to begin. The ghouls would soon descend.

Unless, of course, the three musketeers could find a way to stop them. An epic battle was about to be joined.

Pinecone set down two frothing schooners of beer on the table, then slid into the circular booth beside his colleague Mike Phillips, the second member of the team.

"It's five straight up," said Pinecone. He laid down some bills and silver coins and pushed them across the tabletop to his companion. "Steve should be here any minute."

Mike nodded. He scooped up the change without counting it and slipped it into his pocket.

The "Steve" was Steve Gaylord, the splinter's stealthy third leg, who had been dispatched earlier to keep an eye on the logging site. Very early that morning, they had received a helpful tip from an unnamed source over at the district office that a Forest Service marking crew would leave for Bowen Gulch within the hour, for the purpose of unflagging the dropped unit, the one those three had spiked, and to lay out its replacement. Steve was a born shadow and had been sent to reconnoiter.

Like his eco-mates, Steve had followed his own unique but more or less parallel path to a more radical perspective. Once upon a time in a previous life he had been a computer programmer and a self-described yuppie. But that was in a galaxy long ago and far-far away. Although

Steve's former career had been successful by any measure, he had moved on when the career seemed to him to be going nowhere. The work had come to feel irrelevant, a quaint term from the counterculture days of the 1960s when "relevant" and its opposite expression actually meant something. Both words had been in vogue in that distant past but held little traction, nowadays, and even sounded like the faint detritus of an extinct language. Why persist with a yardstick that was hopelessly out-of-date? Simple. Steve Gaylord was hopelessly out of date, himself.

Indeed, he reveled in the fact and would have it no other way. Steve was a firebrand and something of a misanthrope. He considered it great sport to "get in peoples' faces." How he loved jabbing the jaded ones with pointed reminders that they were leading "lives of quiet desperation." What fun to play the mirror and feed back all of the mindless materialism. The man lived out of the back of his pickup and got by on next to nothing. Despite having few possessions, he flourished, supporting himself on handouts. Not that he was afraid to work. Not at all. Steve was a jack-of-all-trades and worked at odd jobs whenever he could find them, whatever came his way. But gas, food and beer money was all he really needed. Steve Gaylord had found his calling and lived to pursue it. He was a born-again warrior for Mother Earth.

Mike skimmed off the suds as he waited in silence with Pinecone. The beer was good and cold. Although Mike shared the passion of his peers for wild places, he was a different sort of nerd. He was the only one of the three with deep pockets, and thus served as the group's financier. Mike bankrolled the trio's exploits with a roll of cash that never faltered. A new age entrepreneur, he owned and operated a successful chain of sporting gear and outdoor clothing outlets, located in Boulder, Denver and Colorado Springs. The three stores catered to a niche market, extreme sports junkies, skiers, hikers, bikers, wilderness campers, and the like. The stores did well enough. Tourism continues to be the perennial grease for the Colorado economy. Whenever Mike needed to get away he tossed his gear into the back of his 4x4 pickup and headed for the mountains, his release from purgatory.

"Ah, here he comes."

Steve Gaylord slid into the booth with both palms upraised. "Done," he said. "Give me ten." The others slapped back. It was high-fives all around.

"All right!" said Mike.

"The fearsome threesome!"

"They finished," Steve said soberly. The two others fell silent.

"You're shitting me," Pinecone said in a subdued tone that conveyed real surprise. "They *never* move that fast. Never!"

"Well, they did this time."

There had been increasing worry. They had gotten word about a new resolve over at the district office. The timber beasts that still dominated the Forest Service were determined to push ahead with the Bowen Gulch sale, despite mounting opposition from Joe Public. Come hell or high water. It was as if the district ranger and his timber staff took the protests as a personal affront. In the face of growing public opposition to the sale, they were more determined than ever to "get out the cut." As if strip-mining ten million board feet somehow justified their banal existence.

Another long silence.

"Zero hour," Steve finally said.

"Yeah," said Pinecone. "We can't afford to wait any longer. We are out of time. We'll have to move tonight."

"We can't allow them to push the road in."

"No."

"It won't be easy," said Mike.

"Fuck easy," said Steve. "Whatever it takes."

"Exactly" said Pinecone.

"OK. That decides it," said Mike. "Let's go get some dinner and plan the action..." The others nodded. Mike put his clenched fist on the table. "One for all!" he said.

The others placed their hands on his.

"And all for one!"

Thieves were never thicker than these three.

TWENTY SIX

Tallie was brushing Luther down when Mary came into the barn.

"Did you have a good ride?"

"Always"

Mary had an apple in her hand which she now offered to Luther. The horse plucked it from her fingers. "I'm so proud of this guy," she said, stroking his neck. The apple disappeared in two quick bites.

"He's the best."

"Bernard is so thrilled to be able to ride him again. It's the darndest thing I've ever seen. I guess miracles do happen. Luther's made a believer out of me." Mary scratched the top of Luther's head. The animal lifted his nose and swished his tail, loving the attention.

"We healed him," Tallie said. "Tom and I."

Mary looked stunned.

"How I don't know, but we did. I think it happened when we rode him, together."

"It was you two, then?"

Tallie nodded. "I think so."

"So what are you, a couple of horse whisperers?" Mary was laughing. "I was not aware that you and Tom went riding. For what it's worth, I like your new boy friend. Does he know what happened?"

"Not yet."

"You should tell him."

Tallie had pushed her sunglasses up over her forehead. "I plan to."

"That's some powerful energy happening between you. I'm so happy for you." There was love in Mary's eyes.

Tallie was immensely pleased and resumed brushing.

"He's so different from Jake," Mary said. "And thank heaven for that."

Her aunt was right as usual. The marriage with Jake Flaherty had gone wrong almost from the day they tied the knot. The tumultuous relationship had lasted only six months, most of which time they lived apart. It had been one of the most stressful times in her life. Despite appearances, Tallie was actually six years older than Tom. Her stormy second marriage had been much like the first with that boring businessman Harold. Twice married and twice divorced, from men who believed they owned her and sought absolute control.

But Tom was different. From the very first, their coupling had been so easy and right. During the long bus ride cross-country to San Francisco her reverie never faded. Weeks later, she still felt him deep inside of her. Their bodies seemed perfectly attuned. The wild friction of their lovemaking was like silent music, long improvisations of surprise and wonder, free form accompaniments and endless harmonious variations, beautiful music that only they could hear, as they slowly ascended to the summit of their full measure, their dueling bows bent on mutual annihilation. After, she could never remember the things he whispered in her ear, nor even her own solemn benedictions. As if their love words belonged not to this world but to an eternal moment.

Tom made the other men in her life seem like boys by comparison. He was the key to her lock, the hand in her glove. Despite the pain and background noise that usually roiled her head, she had been having dreams of late, waking dreams more vivid than day, about a life and shared things she had never dared to believe might come true.

He had passed her a few lines hastily scribbled on a scrap of paper, a spur of the moment thing...

You stole my heart when you went away.
How I resented your being gay.

But I'm sorry, I was wrong, and pray
For your forgiveness. Suffice to say
I love you anyway.

The poem wasn't much, only a ditty. But it meant the world to her.

"Tom writes poetry."

"Ahhh. I'd love to read it."

She often discussed her intimacies with her aunt, girlfriends as well as men, but there were a few things she did not share.

As Mary turned to go she said, "Oh by the way, we eat in half an hour."

"Good. I am sooo hungry."

"Don't ask me 'what's for dinner?' Bernard won't tell. He wants to surprise us." Mary made a face, mimicking her husband. "The old coot's on the rampage again. He won't even let me in the kitchen. Something garlicky, though. Anyway, it smells good. It better be or I'll divorce him."

Tallie laughed.

Mary was nearly out of the barn when she turned back. "Would you believe I came out here to call you to dinner, then almost forgot?"

Tallie smiled. But when Mary left, her eyes filled with tears.

TWENTY SEVEN

Jacques St. Clair had been cruising east on I-70 for nearly an hour. After crossing Berthoud Pass he had descended through Idaho Springs, and was now fast approaching the western outskirts of the Mile High City. The evening commute was winding down. What remained of the heavy traffic was moving in the opposite direction – out of town. Jacques had entered the notorious smudge of brown haze that hangs like death over the creeping Front Range megalopolis. His business occasionally brought him to Denver. It was an OK city, as cities went...

Just so I don't have to live there.

The boss knew he was fortunate to live and work in the mountains where there was plenty of elbow-room, and fresh air. He hardly noticed the smog and traffic, but his head was a welter of worries and what ifs.

The boss was still fuming about that dandified company attorney, Ted Roe, who had foxed him, earlier in the day. For a languorous moment Jacques rubbed his temple and mentally reviewed the irate meeting on the landing. The conversation with the lawyer had not gone well. Roe personally exhibited many of the qualities that Jacques most despised in a man. He found the snide attitude especially infuriating. Roe wielded the power of insinuation like a weapon.

He thinks 'cause he went to Haaarvad his shit don't stink.

Lawyers! How he hated the bastards and Roe was so typical of the breed. The man was a dandy, alright, with that groomed-to-perfection look, the dapper suit (under his trench coat) and loud paisley silk tie, the designer dress shoes, the cologne, yes, even the gaudy rings on his dainty fingers, and Christ, those uncalloused hands! The man obviously had never done a lick of physical labor in all of his miserable existence.

As Jacques relived the rancorous confrontation he began to rage all over again. He could never understand why it always seemed that honest working men like himself were lorded over by assholes like Roe, men who lived off the sweat of others and never produced a goddamn thing themselves of any value, only trouble for others. Such men were no better than parasites. Jacques recalled the sneer on Roe's face when he slipped him the Mickey.

That lowdown snake...

Western-Pacific had been awarded the contract for the Bowen Gulch timber sale. However, it was standard for the big outfits to sub out the actual work to smaller operators. Jacques disliked doing business with them because he resented the interference. He preferred to run his own show and he did, whenever he could scare up his own jobs. He and his brother were always bidding on contracts. Bowen Gulch, however, had been too tempting to pass up. The unprecedented size of the timber translated into *big* money for all concerned. Now, belatedly, Jacques wished that he had heeded his brother Paul's advice. His older brother was his business partner and half-owner of their Leadville-based outfit, Right of Way Inc. Earlier that spring Paul had advised him in the strongest language to review the sub-contract line by line with Tim Hollinger, their private attorney. As usual, though, Jacques had been in too much of a damn hurry.

It was now painfully evident that the fine print and his haste earlier that spring might well end up costing him. It all depended on how far W-P was prepared to go. At the moment, his top priority was damage control. Before leaving Granby, he had attempted to reach Hollinger by phone to discuss the escape clause. Unfortunately, his attorney was out of the office. He would try again in the morning.

Damnation!

A man could wind up in the poor house overlooking the subtleties of contractual language. Was it already too late? He wondered. The clause loomed like Everest. He wondered if he had been "set up" by company lawyers who at this very moment were probably snickering in their double martinis.

The possibility made his blood boil.

Jacques struggled to clear his head. He knew he would have to get a grip to salvage the situation. He began to review in his head the list of things he needed to accomplish, next day. It was a chore just sorting and prioritizing.

Number two on the list was to arrange rental terms for two replacement skidders, assuming he could locate a dealer. Before leaving Granby, he had phoned his usual contact in Denver, only to receive more bad news. The agent presently had no skidders available. Worse, the man did not know of any other dealers in Denver who did. Jacques would have to shop around.

There was also a third pot-boiler, regarding payroll withdrawals for workmen's compensation and state disability insurance. The phrase used by the state in its recent letter was "certain irregularities," words with an ominous ring. For weeks, Jacques had put off the matter. Dealing with state finaglers was far down his list, certainly not his favorite pastime, but since he was going to be in Denver, anyway, he had put through a call to the state office and set up an appointment. After some haggling, they agreed to meet with him, on short notice. The conference was set for 9:00 A.M. in the morning. *How that* would go was anybody's guess.

The boss dreaded the rendezvous with state bureaucrats even more than his troubles with W-P.

Based on past experience, Jacques was convinced that the state bureaucrats were a bunch of cynical bastards. How they loved to dicker with a man. The bureaucrats could drive an operator to drink with their high-handed attitude. The rules changed so often that a man could never be sure, from one month to the next, whether he was still in business or in hock. They administered the rules so arbitrarily that you never knew where you stood. And there was no appeal, the state boys were classic turf-lords, a fiefdom unto themselves. Answering to no one, they took positive delight in burying a small operator under a mountain of regulations and red tape, that is, when they were not taxing him to death, driving him into chapter eleven. As if the IRS were not bad enough.

If it isn't one thing, it's another...

For five years running, Right of Way Inc. had been audited by federal revenuers. To protect themselves Jacques and his brother Paul had been forced to hire a part-time accountant, even though they believed that the size of their company did not warrant such an outlay. Five years in a row the accountant's fees had nearly matched their federal tax. Not to mention the inconvenience and the increasing portion of their time and energy that was consumed with federal bullshit; whether filling out endless forms in triplicate or making redundant phone calls.

This sort of busy work always gave Jacques a helpless feeling, as if he were drowning in a sea of red ink.

Well ... Isn't it the truth!?

At times, he suspected that the feds and state bureaucrats were in cahoots. In the end, what could a working-man do about their legal chicanery?

Not a thing! Except write your crooked congressman and, in the meantime, bellyache while they slowly bleed us to death.

The anxiety often affected Jacques, in the grotto of his nervous stomach. His sense of helplessness was usually followed by mounting rage.

Oh why can't they leave an honest man alone so he can make a living?

Never before in his years in the woods had St. Clair faced a bottom line worst-case scenario on a job. One way or the other he had always turned a profit. The realization that Bowen Gulch might become a debacle and, quite possibly, the first major financial setback of his career, now gave him pause. Jacques shuddered as he visualized his profit margin slipping away, yes, like an August snowfield.

Merde!

Denver traffic brought him back. He steered his rig into the right-turn lane and entered the cloverleaf interchange with I-25. He was headed south into the heart of darkness – downtown.

Dusk was still a half-hour away. But the sodden sun had already slipped into the brown sludge that passed for a western sky. The Front Range was also gone from sight, concealed by the noxious crud. Lights were winking on across the Mile High skyline. The traffic was bumper-to-bumper.

For a moment Jacques closed his eyes and recited a silent prayer, the last part of which he modified, or rather improved with some untranslatable French.

Oh and there was another troubling matter the boss would have preferred to expunge from his thoughts; but somehow could not. The nagging possibility that he might be held up in Denver longer than expected, and be unable to return to Bowen Gulch for the start up. If he ran into any unforeseen problems in the morning...

That would not be good. But let's not think about that, yet.

TWENTY EIGHT

A lone dot circles in the vault of sky. Soundlessly. For several seconds it disappears, lost in the blue vastness. There it is again, a winged phantom on the silent edge of a long and lazy arc. It's a Red-tailed hawk, soaring skyward on a late afternoon thermal. The raptor's flight is graceful, effortless. It floats as if suspended, as indeed it is, on a rising column of warm air, a sky-bound elevator. Now, its wings bob and dip, turning and wheeling about, feathers ruffling in the breeze.

The hawk catches a fresh updraft and rises fast; but its eyes remain grounded on earth far below. The hawk is on the prowl.

From this great height the view is unobstructed for many miles in every direction. Two rugged mountain ranges dominate the landscape, each glistening under a snowy white mantle. To the east is the imperious white wall of the Front Range, several of its peaks clawing more than 14,000 feet into the sky. Nearby, just ahead, lies the jumbled spine of the Never Summer Mountains, the southern extension of the Medicine Bows.

Now, the aerial hunter drifts laterally and generally southward, crossing ridge after ridge. As it does, its sharp eyes scan one alpine valley after another. Each of the forested valleys is a patchwork of unnatural disturbances. Some of the scars look recent. Several were not in evidence a few weeks before, when the hawk last passed this way on its semi-regular hunting rounds. In this high mountain fastness, where hundreds of feeder streams swollen by snowmelt come crashing down from the surrounding heights to nurse the great rivers of the continent, conifer green is giving way to ruddy brown.

Presently the hawk hears a faraway noise carried on the late afternoon breeze. It's a low mechanical sound, faint, a distant low rumble. But the sharp-eyed hawk has no difficulty identifying the source. Far below, nearly a mile away, a tiny yellow machine crawls like a bug along the margin of an eviscerated forest. The yellow object slips from sight, hidden by a fringe of trees, then reappears and emits a tiny puff of black smoke seemingly disconnected from the sound.

Circling, the hawk studies the strange object. After awhile though, it loses interest and moves on.

Heading east, the Red-tail leaves the scars behind and passes over the north-south backbone of the Never Summers. The hawk has crossed the great divide and now soars above a wide green valley, the headwaters of the Colorado River. The valley is flanked by massive escarpments, U-shaped lesser valleys and ridges fanning out from the snowy heights. Gentler landforms attenuate the harshness of rocky uplift and ice.

The floor of the great valley is mantled by a coniferous forest that stretches unbroken, save for a few scattered lakes, occasional meadows, and outcrops. The forest remains much as the Creator made it, nearly untouched by human hands.

Presently, the Red-tail loses elevation as it glides toward its destination, a meadow not far from the great river where on innumerable previous forays the hawk has hunted up its supper. On the edge of the meadow stands a large white snag, an excellent high perch on whose skeletal limbs the hunting bird has often patiently roosted, before swooping down to snatch a tasty meal of chipmunk, ground squirrel, or jack-rabbit.

This day, however, the raptor finds that intruders have occupied its familiar hunting grounds. Intrigued, it circles, aerially detached, yet absorbed, as it studies the strange objects beneath the trees. They are alien to the native rhythms of this place.

Though fascinated, the hawk is not pleased by this rude entry, which has frightened away the usual game from the vicinity. The Red-tail will have to revisit other haunts before the day is through to scare up food for its hungry brood; which will mean expending more energy logging extra miles on the wing. Oh bother. The hawk turns and starts to drift down valley.

But wait.

It arcs again and spirals back. Strange pedestrian critters are moving below, hidden for the most part under cover of trees. Occasionally, one or two of them pop into view. Aha! Recognition! The Red-tail has seen these critters on many occasions and, as always, is fascinated by them. What a clumsy means of locomotion. What an awkward way to get from one place to another, staggering about like storks on ungainly hind limbs. Too bad these fur-less varmints are too big to bag. Were they a bit smaller they would be easy to catch. One of them would feed the entire brood.

But what's this?

The hawk wheels about, circling with renewed interest. Something unusual has grabbed its attention. The creatures appear to be assembling. Six. Seven. Eight. More are coming. It's some type of social gathering. What strange creatures they are. But what exactly are they up to?

TWENTY NINE

Dusk, Kawuneechee Valley

The men crowded around Tom's cook fire for a humongous spaghetti feed. He had put out the word about the free meal, and when the chow was almost ready sent Pissant around camp banging on a cook pot to announce that s-s-supper was b-b-being s-s-s-served!

Almost everyone showed up, except the Preacher and Bobby Whitehorse, who thought it best to remain out of sight.

The men were not about to pass up a free meal and they brought their hunger with them. Loggers can really pack it in. The planners had hoped for as much, and made certain there was plenty to go around.

The dinner did not disappoint. The spaghetti sauce especially was a hit, and registered high praise from the men. "Mmm, great sauce," said Sourpuss, gesturing with his fork. "What'd you put in it? Tastes kind of like mushrooms..."

"That's right," Tom said. "Fresh and locally picked."

"Well, you done good. It's dee-li-shus."

"And good for you too. They are loaded with anti-oxidants."

"Anti-what?" said Wolfe, sucking a noodle.

"Think vitamins," Tom said. "Nature's very own." Wolfe just grunted and dipped his snout deeper in the trough. Free food met his approval under any circumstances.

Francis Delacour was a late arrival. There was some doubt whether he would show up because, although on Jacques' orders he had parked his camper trailer at the logging camp with the crew, Francis had been spending every spare moment in town with his wife whose time was very near. The sabotage of course had changed everything. Before leaving for Denver, Jacques had charged Francis "to run a tight ship while I'm gone." That meant spending his nights at the logging camp where he could keep an eye on the dozer and also be closer to the rest of the equipment at Bowen Gulch, which was about three miles away, up a rough logging road.

Francis shared the good news. "The flagging is done, boys," he told them. "We start tomorrow morning, bright and early."

"Hey wow, Bowen Gulch. I heard there's some big-ass timber in there. We finally made the big time."

"Yeah," Dipstick crowed. "We gonna' rake it in. After this job I'm going south to Mexico."

"At's right. Take the money and run."

The Forest Service had done the near impossible. Incredibly, the marking crew had laid out the new shelterwood unit, both the marking and flagging, in a single day. Such rare efficiency was unfathomable.

"A minor miracle," Francis said, "that we owe to Bill Noonan. He and the ranger ordered everyone out of the office to help, even the secretaries. The staff put in a twelve-hour day and got her done last night, sometime after dark. Noonan lit a fire under their candy asses." This elicited laughter.

Francis also filled them in about the meeting with the company attorney, and how he had T-boned Jacques. "The boss thought he had a contract," said Francis. "But the company lawyer blind-sided him with that escape clause. After the asshole left, though, Jacques kind of regrouped. The last thing I heard him say was: 'Screw the bastards'!"

This drew smiles and nods of approval from the men.

At that moment, Delacour's pager went off. "Oh hey!" he said. "That's Rose. I got to go." His wife apparently was starting into labor. Francis had just finished loading up his plate but set it down again in a hurry. He had not taken a bite. "I'll grab something at the hospital."

Before he left for Granby, though, Francis pulled Shorty aside. "See that dozer over there, Shorty?" He had him by the shirt.

Shorty nodded.

"I want you to guard it with your life. Understand?"

"OK."

Francis hesitated, evidently not certain that he had gotten through to the man. Or maybe he just wanted to avoid a miscommunication. Shorty, after all, was not the brightest light in the West. "Repeat after me," Francis said, "It'll be my ass..."

"It'll be *your* ass..."

Francis fingered him hard in the ribs. "Listen, wood-for-brains, if anything happens to that cat it will be *your* butt. Are we clear?"

"Clear, Francis, clear."

"OK. Good."

Francis turned to leave and said with a dismissive wave, "See all of you birds in the morning."

"Night, Francis."

Shorty got the message. A few minutes later he climbed up on the dozer with his chow in hand and ate his spaghetti dinner in the catbird's seat.

The crew put away every last morsel, down to the last noodle. Not even a spoonful of sauce remained. Tom noticed Wolfe licking his fingers after wiping the pan. When the men were done, and well satisfied too, they urrped and belched their way back to their trailers.

At that point, the logging camp settled down. Things got quiet and stayed that way for the next hour.

Earlier, Tom had discussed his reservations about the mushroom caper with Bobby. It seemed a broad net just to snare one fish. "What does that make the rest of us, guinea pigs? Or collateral damage?"

"Hey, now, friend. You are thinking about this in the wrong way. All will be well. Trust me."

Trust, however, was in short supply. "You're kidding me, right?"

"I thought you said you did acid in college?"

"Yeah."

"Well these mushrooms are bokou superior to LSD. They are wholly organic. Smooth as smooth can be. The cream of Mother Nature's own sweet brew, ambrosia of the gods. As for the rest of the crew, why are you worried? These guys are about as hung down and wrung down as a man can get. There's not one of them who won't benefit from having his world-view expanded and his consciousness raised. Would you not agree?" True enough, Tom had to admit. But Bobby was not done yet. "Heck, their minds run in narrow grooves. They are completely predictable, creatures of habit. They talk in clichés. They think in stereotypes. Worst of all, they are boring. B-O-R-I-N-G. Sometimes I can hardly stand their stale crappola." It was all true, every last word, so very true. "Did I miss anything? Did I leave anything out?"

"Nope. You covered it."

"The worst thing about being hung down is you don't even know just how wrung down you are."

"Right. Agreed."

"How can you think 'out of the box' if you don't even know you are in one? You don't. You can't. That's why we are doing these fellows a big favor. They need to have their minds blown, and opened up to a bigger reality. A few dozen of these little mushrooms should do the trick. I intend to indulge myself. Have no fear, brother. This manna came straight from heaven. Medicine for whatever ails you."

"If it's true, Amen to that."

"It is oh so true, brother. Trust in the crazy wisdom of the Great Spirit."

"OK. I'm just crazy enough to try anything once."

"Now you're talking."

But that was not quite the end of it either. Tom also had doubts about whether the active chemicals in the mushroom would survive the heat of the cook fire. It was decided to only lightly brown the mushrooms in the interest of potency, and to fold them into the sauce just before serving. Bobby contended that even if the effects were diminished by half, it would make little difference given the mushroom's incredible power. The matter was not finally resolved until holy hell broke loose.

The next thing anyone knew, Charlie McCoy was thundering through camp. Night had fallen.

"I'm mean you!" Charlie bellowed. He had planted himself outside Wolfe's camper and was calling the man out. "We've had enough of your low life bullshit!" cried Charlie. "You sick fuck! We think you're light in the stones department! Hey, I'm talking to you in there!"

Attracted by the commotion, the loggers now gathered around. Charlie was ready for bear, his fists clenched. There was no sign of Wolfe though, until finally his head appeared in the trailer window.

"We'll settle this, today. Here and now!" Charlie roared. Evidently he had taken upon himself the unpleasant chore of ridding camp of this no good ex-con whom everyone agreed was trouble walking. It had come to this.

Wolfe spoke not a word as he stepped down out of his trailer, his face a mask, betraying nothing, no hint of feeling, raw malice in his eyes. Tom shuddered at the sight of him. It was as the man had fallen out of an ugly tree and hit every branch on the way down. Tom had an intuition that Bobby was right.

He's a cold-blooded killer alright.

"You and what army?" Wolfe whispered through his teeth. He was taller than Charlie by several inches, and outweighed him by at least twenty-five pounds. Yet, in recent days there had been talk about camp that if anybody could take Wolfe it was probably Charlie. Shorty was another possible candidate. But Shorty was streaky, at best, when it came to a fight. Charlie was a former Navy boxer and a lot tougher than he looked.

The two men circled in and out of the light from the camper window. The outcome was up for grabs. The crew silently waited with bated breath. It was anybody's guess which gladiator would get in the first lick.

"I've been looking forward to this for a long time," Charlie said, rubbing his nose with his thumb, then, he took the first swing. But he was wide and Wolfe clipped him a good one on the forehead. Charlie fell back but recovered and came in again swinging, his feet dancing. Wolfe fended off the punch and pushed him away. The muscles on Wolfe's neck were taut cords.

"Is that the best you got?" he taunted. "My scrawny little sister hits harder than that."

They circled around some more and Charlie made another move, but not with his fists. He bull-rushed him, came in low and caught Wolfe about waist high. The two grappled as Wolfe pounded on Charlie's back. They went down together, rolling over. But they separated with astonishing speed and were back on their feet. Things had gotten serious. There was no more talking as they circled quietly, each probing with jabs, waiting for an opening.

Wolfe swung next, but Charlie stepped under the punch and delivered a right upper-cut to the chin that landed with a solid thud. Wolfe looked stunned, his eyes blank. A tremor passed through him and he sagged to his knees. The fight was over. Everyone knew it.

"Sock on baby!" someone cried.

Charlie moved in to finish him. In mid-swing, however, he pulled up, a surprised look on his face, a strange glimmer in his eyes. The next moment Charlie relaxed his fists and started laughing. Incredibly, Wolfe was laughing too. They were laughing at one another. "I win, and that's that!" said Charlie.

"Yeah? Well, you better finish me, 'cause I plan to do your mother!" Wolfe cried. "Ha-ha-ha!"

But Charlie just laughed along and began to taunt Wolfe about his ugly sister.

It was insane. The two men kept hurling insults at one another, each one more outrageous than the last. But the shit would not stick. By now, they were laughing too hard to fight any more.

Whatever had happened was apparently contagious, because the uproarious laughter now spread to the rest. It was as if a breath of enchantment had escaped suddenly from a bottle; and that is how Jacques St. Clair's loggers came under the sway of the mushroom.

A person with a sober eye would have concluded that their minds had been carried off. But what do sober people know?

That was when the Preacher showed up. He had missed the communal meal and now stood ashen-faced, shocked to apoplexy by the wacked-out craziness assailing his senses from every side. Tom had never seen a look of such dismay. But the man's shock quickly morphed into a religious frenzy. Enraged by the scene of unChristian debauchery, the Preacher began spewing his personal brand of fire and brimstone at them, meanwhile, wagging his finger of doom. But Dipstick had come up behind and now popped the Preacher's suspenders. As he hurled his best shot, "a terrible wrath, the judgement of heaven, is coming and will soon befall you," his baggy trousers dropped to his ankles.

The howling that ensued would be unrepeatable to polite company. Nonetheless, I will do my best to give an accounting...

THIRTY

Something was happening. Everything was out of joint and seemed ridiculous. A feeling of lightness, inner space, distance. Mostly, just a profound silliness. Charlie and Wolfe were still going at it, screaming and laughing at each other. But Tom noticed that their words were not connecting somehow, like they were coming from afar. He watched Charlie's lips moving. But the voice was out of sync. The words were like silent meteors streaking by, flashes of sound. Not that he *couldn't* connect. He found that if he tried, if he really concentrated on the chore of listening to what the men were saying, he could follow everything. The thing is, *he didn't much want to.* He didn't *feel* like listening? Listen? Why bother? The very idea was becoming more absurd by the second. What was happening inside, what he was feeling was much more interesting than language. Words? What were they? So much wasted breath. Much ado about nothing. He chuckled at Wolfe and Charlie, and wondered if they and the others were also feeling what he was feeling.

Bobby had described how the mushroom came on gradually, sneaking up, then, "Wham! It hits you like a freight train. Between the eyes."

Yep, wow, that's how it is. Like a freight train coming on hard and fast.

Tom's eyes were wide as spoons.

The feeling of being in the present was intense. There was no sense of the passage of time. He felt perfectly attuned, centered in his body. There was a physical dimension to it and in this respect the mushroom was different from most of the hallucinogenic experiences of his college days. During his junior year at the university he had experimented with psychedelics. He tried windowpane, purple haze and blotter acid, and once had a bum trip on synthetic mescaline. Another time he also did peyote, a truly memorable trip. Meanwhile, he avoided the hard drugs, amphetamines, downers, and tranquilizers. None of them held any appeal. He was not into escape. He was into exploration, consciousness, pushing back the frontiers, opening the "doors of perception."

But none of those experiences were like *this.* This was qualitatively different. So far, there were no rainbow effects, no flashing after-images, nor any chromatic wonders as with peyote, and thankfully, none of the nausea. There was no attenuation of time either, as with LSD, in other words, no watch hands moving backwards. There was no weird sensory distortion, no swarming bugs, and thank heaven for that. There was no hard metallic edge, no shakiness, and thankfully, no diaphragmatic distress, none of the itchy-crawly needly feeling of flesh-on-fire indicative of chemical adulterates or histaminic impurities. There were no jitters and no shakes. Whatever the mushroom's active agent or agents were, the effect was silky smooth and wonderful, just as Bobby had described it.

The high had a natural organic feel to it. There was a powerful sense of the wholeness of Nature, a strong awareness of the harmony and unity of life. The experience was ineffable. It had a cosmic quality, utterly beyond the power of words to describe. In this respect, the trip was similar to peyote, but minus the nausea and rainbow color show, just the humming of a body/mind full unto itself, dancing in step with the universe.

As the mushroom high came on stronger, his eyes started buzzing, then his jaw. A flood of energy was washing through him, chemo-electricity, liquid love dancing and pulsing through every artery, nerve, muscle, bone, follicle, cell, and pore. He was riding a wave of pure energy. He felt that he finally knew what the poet William Blake was talking about when he wrote that energy is "eternal delight."

Now, however, he encountered a snag. A lone doubt emerged and grew into a difficult moment. An impasse. He knew he was in a process of mind expansion, losing his ego. But a part of him now resisted going there. He felt smothered. One of the men had come up and was giving him a big bear hug. He felt his breath being squeezed out of him. Suddenly, he was paranoid, and for the first time knew fear. For a timeless moment he fought the mushroom, struggling with himself, on the verge of disintegration. He recalled what Bobby had told him. "Losing your ego can be scary, like death," he'd said. "Let the barriers come down. Let it all go, brother; don't hold anything back. We are all one, your skin, my skin. Get past the panic and it'll be smooth sailing. Remember, we all go through it."

Trust in the crazy wisdom...

He actually saw himself hefting his old hoedad, swinging it with practiced ease, first up and then down like a vajra blade, slicing through all barriers. That did it. Something let go and suddenly he was in the flow again, riding a swift whitewater current downstream.

Now, wave after wave swept over him and he began to merge with the energy. Gradually, imperceptibly, the distinction between "inside" and "outside" lost all meaning. Incredibly, he also understood why. When you got right down to it, the distinction between "I" and "other" had no intrinsic philosophical validity whatsoever. Was this not self-evident? Ego? What was that? Nothing but an arbitrary mental boundary, a social convenience, a perceptual contrivance.

By now he was no longer riding a wave of pure delight. He and the wave had fused and become one. Suddenly the answers to questions that had preoccupied philosophers for centuries seemed perfectly obvious. Cartesian dualism? "I think, therefore I am." He laughed at the absurdity of such an epistemological formulation. Never had human thought conceived a more shallow and arbitrary philosophical first principle. Descartes had imposed an ego where none could be shown to exist. Why? Because the only first statement that could justifiably be made was, "There is a thought." Yes, David Hume had been absolutely correct about that.

And how strange it was that a philosophical starting-point as naive as dualism had ever gained such wide currency; at least, in the West. It was, he decided, a sad commentary on European thought. For thousands of years Buddhists and Hindus had known better. In the West the matter had not been dissected properly until David Hume did it, building on the work of Locke and Berkeley before him. A most amiable corpulence, Hume. As a philosopher he also possessed the virtue of great modesty. To his credit, the brilliant skeptic correctly insisted on the primacy of direct experience. At this point, however, an intriguing consideration arose regarding Hume's famous repudiation of any logical connection between cause and effect. This was another kettle. At issue was whether Hume would (or could) have defended his position in the midst of an experience as ineffable as *this.* The answer again was self-evident. Gleefully Tom shouted it, "Of course not. Never!"

No philosophy dependent on linguistic distinctions could stand in the face of this seamless and irreducible whole that slipped through the cracks of language like water through a leaky vessel. When you got down to them, the answers were simple. Human language was a sieve. A single mushroom would have exploded all of Hume's doubts. And what of Kant? One such experience would have exposed his *a priori/a posteriori* reasoning for the linguistic contrivance it was. In matters of inquiry, what greater authority could there be than this rising tide that kept building within (and without), buoying a man's spirit as it lifted him higher and higher?

Tom was now ecstatic. Happiness welled up within and without him, bubbling over limitlessly. Tears of contentment flowed from his eyes. Jubilant, he wanted to announce his existence to the cosmos, the simple fact that he was alive. He wanted to celebrate, and express the wild exuberance and joy that he felt.

He was surprised to discover that he was lying on the ground, flat on his back. So, first he had to reorient himself. But no problem, he found his body again and rose to his feet. The very concept of "body" seemed hilarious. For what might have been an eternity he stood in the darkness immersed in deep wonder at every sensory impression, marveling at each facet of existence. He had now wandered away from the others, though he could still hear their wild racket, laughter and general craziness.

His eyes had adapted to the dark, and he skirted several trailers and became absorbed by other sounds. Melodious frogs were croaking down along the creek. They might have been angels singing, or Mozart communing with the Almighty. From somewhere came the double hoot of an owl. Now reaching the road at the rear of camp, he stepped out from under cover of fir trees and aspens. The full splendor of the night sky spread out above him. Any other evening he would have approached the starry heavens in a rational and methodical manner. He would have started with the Big Dipper, fixing on Polaris, orienting himself, noting the Little Dipper, moving on to Orion, Sirius, the Pleiades, and so forth. But not tonight. All he could do was gape. He was blown away by the totality of it. The Milky Way was brighter than he had ever seen it, splashed across the firmament like a river of multi-colored splendor. Most remarkable of all was the extra dimension, a sense of depth to the night sky that he had never noticed before. Tonight, he was not simply looking up *at* the stars. No, incredibly, he was looking *through* them, and beyond...

"How did I miss this?"

He chuckled at the sound of his voice. Fast on the heels of this came another discovery, the sense of being "up close" to everything. No longer were the stars remote. They had ceased to be impersonal specks of light. No, they were up close, *so very near.* Thousands of light years were as nothing. He felt that he could reach out, yes, and touch the stars. Not quite, but almost. Yes, and there was another novel impression as well, a feeling of...what? Familiarity? *Déjà vu,* perhaps? Kinship? It was a feeling like...?

For a moment he sensed that he was on the verge of something momentous, a great discovery, an insight of historic importance. However, before he could slip his mind around it, pooof! The idea escaped him. Gone. For a disconcerting moment he felt that he would never be able to express, nor remember, the intangible something about this night sky that for just an instant he had sensed so deeply. But he also knew, yes, he was certain that the answer, the intangible "something," lay hidden within himself. In a strange and inexplicable way *He Himself Was the Answer!* Yes, and he knew this with absolute certainty. Somehow, the truth in this night sky had to do with Tom Lacey. What is more, *it was something he had once known, once upon a time.*

Only, I forgot.

What could the "it" be? The insight had been too fleeting and elusive to remember, just a glimmer, impossible to hold. Yet, there had been a quality about it...what? The word that came to mind was "sublime." He laughed at himself. There he was, back to playing with words again. What was the imponderable something that was so important? Impossible to say.

Now the stream of his thoughts ran wild and free. He began to entertain ideas that on any other night he would have dismissed as the ravings of a lunatic. But not this night. No! This was the night of nights! This was a night to howl at the moon and stars. This was a night to be all that a man could be. It seemed perfectly reasonable to speculate about the most far out things he could imagine.

Can it really be?

It seemed to him, now, that all of the great philosophical questions boiled down and reduced to one. Yes, in the end all of the great questions rolled up together. What is more, he felt certain that he was very close to the answer. The "it" was close at hand. Very close indeed. Somehow, he knew that the ultimate meaning that he sought was now within reach. It was right there before him, just waiting to be grasped. He lifted his arm. Suddenly, he knew that the answer lay just beyond his fingertips, out there amidst the galactic stardust of a trillion blazing suns.

He felt more alive and more fully in command of himself than at any time in his life. Anything that he could imagine was plausible simply by virtue of the fact that he could think it. He felt supreme confidence in himself and his abilities. He could do anything that he wished to do, do it with ease, and yes, more proficiently than ever.

All of this time the energy had been mounting. Now, the mushroom began peaking inside Tom's head like a fountain. In the next moment he knew with inexplicable certitude that he was gazing not simply at the stars of the Milky Way, he was peering into the depths of his own soul. He was plumbing a consciousness greater than himself, a reality vastly bigger than the human mind. He had reached the deepest well, the root of universal being...

Yes! But of course!

What were these tiny pinpricks of starlight, anyway, if not the kernels of his own precious thoughts? Here was the root! Yes, the seed! The first kindling spark of awareness in the fertile void of existence. Now, he understood. Finally. At last. The answer was obvious. It was inherent! Self-evident! Implicit! The kernel, the seed, had been there from the beginning of time, only hidden. Now it was revealed. He had found the meaning of the "it" lapping on the shore of...the great sea of being. What else? How profound! What magnificence!

What splendiferous glory!

A final surge took out his synaptic junctions. Somewhere in the mysterious space between Tom's ears a gentle volcano erupted like a tower of translucent light; and Tom, or rather, his essence, was drawn up into the vortex, gently, higher and higher, first to the far planets, then beyond, even to the penultimate stars.

THIRTY ONE

For the last ten minutes Sheriff's deputy Joe Ramirez had been cruising north out of Granby on Highway 34, watching for the turnoff to the Kawuneechee campground. But when he noticed the sign he was going too fast and overshot the turn. Abruptly the deputy hit the brakes, made a "Uee", and went back. He turned and drove cautiously up the gravel access road to the campground proper.

The cab was dark but the red light on his police radio was blinking steadily. Static chattered over the airwaves, several loud spikes in succession. Now, came a familiar but muffled voice. He picked up.

"Is that you, Sherry?"

"Roger, Joe. What is your ten-twenty?"

"You are mostly static," he told her. A large thunder-storm west of Denver was playing havoc with the air-waves. This evening, there had been numerous lightning strikes. "Am just now arriving the campground," he said. "Will keep you posted ASAP what I find. Over."

"Ten-four. Just checking. Will await your call. Over."

Ramirez was not a happy trooper. Earlier, he was about to go off duty when he received an urgent dispatch that wiped out his evening, again. He had plans to go out with his girlfriend, Juanita, which he had been forced to cancel at the last minute.

That's the second time this week.

Earlier in the day the dispatcher at the Sheriff's office had received an irate call from some timber operator. The man filed a complaint that someone, probably environ-mental extremists, had caused extensive damage to his heavy equipment. Anyway, so he claimed. The timber boss was convinced that another attack was imminent, and had demanded immediate protection.

Dispatch promptly sent out a patrol car to the logging site, reportedly in the Kawuneechee Valley near the National Park boundary. However, the National Forest was an impossible maze of secondary and back roads and the officer had been unable to locate the site of the disturbance. They were still trying to check out the story. The timber boss apparently was somewhere in transit. At the moment his whereabouts were unknown. They had not been able to reach him again to ascertain more details. However, the campground, at least, was on the map. So, Ramirez had been instructed to check it out. The order was to secure the site and if necessary lock it down, in order to prevent or respond to any criminal activity.

Joe Ramirez was resentful and felt targeted himself. Why? Simple. He was of Mexican-American ancestry. Prejudice was rife in the force. It was something he lived with, but had never gotten used to. Joe was certain he never would. He hated being the low man on the pole. Even so, it was pointless to complain. That only made things worse.

Such were his thoughts as Ramirez reached the camp. As he swung his car around a number of camper-trailers flashed in the sweep of his headlights. He pulled alongside one, and opened the door of the patrol car. Quickly he checked his gear. He made sure his piece was loaded, standard protocol, and grabbed his Maglite. The night was pitch black but for the flickering light of a campfire. He thought he heard voices, but was not sure.

Moving toward the fire, where the noise seemed to be coming from, he scanned with the light as he cleared the back end of the first camper. But the training academy and his four years as a trooper had not prepared him for what happened next.

As he swept his flashlight he noticed a figure to his right; a blur in the dark. It appeared to be a large man and, here was the strange part, just sitting in the darkness. Instinctively Ramirez went into a crouch and reached for his revolver. Unsnapping the flap, he gripped his weapon with his right hand. But almost immediately he knew he had over-reacted. Loosening his grip on the .38, he stood up and moved closer. It was a large man and he appeared to be on his knees. Or maybe he was just sitting on the ground. From this distance Ramirez could not be sure. As he moved closer he put the beam directly on the man, but the guy never looked up and continued doing whatever he was doing.

Ramirez was startled. It was unusual that a man would fail to acknowledge the arrival of law enforcement. The person looked to be intoxicated, which probably explained his behavior, but Ramirez was not certain. He could not see what the man was doing. He moved nearer and was startled again. The man had his cheek against a tree and ... was stroking the bark with his hand...

What in hell...?

The deputy had no idea. To all appearances the man was totally absorbed with the tree. That was all. That was it. A shiver passed through the deputy.

Yes, it was strange. But Ramirez sensed no threat so he moved on. Now, he swept his light across the flat surface of a camper trailer and noticed it was covered with Day-Glo. He concentrated the beam. There were swirling red bulls' eyes, spirals, all manner of graffiti. Was it a prank? Or something more serious, evidence of sabotage by tree huggers?

He moved toward the fire and nearly tripped over another body, a man lying on the ground. Passed out? Or dead?

"Shit."

He shone the beam on him, up and down.

What...?

More Day-Glo. The man's face was blue. But he wasn't dead, his lips were moving and his eyes were open, wide open. The guy was ... whispering to himself.

Too weird.

The man appeared to be inebriated. The deputy thought so, but he was not sure. He could not get a handle on what he was seeing. During his years on the force he had run into plenty of drunks and druggees, but nothing like this. There was something about this drunk that did not seem right. He made another closer pass with the beam, studying the face. The man was smiling in the strangest way, almost like...

Ramirez moved on. Two men, no, three, half in shadow, were sitting around the dying fire. Three flickering shapes. He spotlighted them with the Maglite, one after another.

Un-fucking believable.

It was a circle jerk.

And I thought I'd seen everything. Three grown men.

Sweeping the light more widely, now, he noticed another big man sitting on a nearby tractor. The piece of heavy equipment was also covered with nonsensical graffiti. Spirals, figures, circles, what-all. Nonsense. The guy was clearly wide-awake and also had a Day-Glo face. He was not doing anything. The man was just staring into the darkness...

Ramirez steadied the beam. There was nothing threatening about him. The guy was peaceful like the others, with the same spacy smile.

As the deputy strode through the darkness back to the patrol car he tried to make sense of it all. But none of it added up. For a moment he half-seriously entertained the possibility that someone was playing a practical joke on him, just one of a many disjointed thoughts and impressions racing through his head. By the time he reached the vehicle he was convinced the report about sabotaged equipment had been a mistake, or maybe a hoax.

He reached in the car door window and grabbed the mike. The curly-Q cord stretched out like a slinky.

"Unit 3-0-5 to base. Do you read?"

"Roger. Is that you Joe? You're coming through clear, now. "

"Copy, Sher." The deputy hesitated. He was unsure how to begin. Suddenly, he wished he had delayed the call for half a minute, to collect his thoughts. It was all nuts, just nuts.

"Joe, I lost you. Are you there?"

"Roger that. Sorry...Look, Sher, uh..We got a situation, up here. OK? Get a grip. You probably won't believe what I'm going to tell you." He paused. "Wait. What was *that*?"

THIRTY TWO

It was after 10:00 P.M. when Jacques St. Clair pulled into the Rustic Resort, half way up the Poudre River road. He was exhausted after an emotionally grueling day. He had spent the morning in Denver sequestered with state bureaucrats who informed him that Right of Way Inc. was currently four months in arrears to the State Workmen's Compensation Insurance Fund for disability premiums. Although Jacques hotly disputed the charge, he insisted it was all a misunderstanding, formal notice was served that the state was about to enforce compliance by slapping a lien on property owned by the company. His dander up, Jacques spent the better part of the next two hours playing phone tag, first, with Priscilla his secretary, then, with his accountant. When he finally stormed out of the state offices around noon the matter still was not resolved. At that point, all Jacques wanted was to find a saloon on Larimer Street and drown his troubles in an anonymous pitcher of golden Coors.

He did not do it, of course, though he was sorely tempted. He still had two replacement skidders to run down. Not to mention that stupid escape clause. So far, he had not been able to contact Tim Hollinger.

There followed another round of calls to dealers in the Denver area. It took longer than expected to run down even one skidder. After calling around, he finally learned from a friend in Cherry Creek about an operator based in La Porte, just north of Fort Collins. The man had one skidder for lease, including the driver. Although Jacques needed two, he would have to settle for one. He closed the deal over the phone, then made haste to get out of Denver before the start of the afternoon rush hour. It was a relief to put that brown excrescence that passed for Denver air behind him.

Phewww...

He never did reach his attorney.

As he was cruising north on I-25 Jacques finally realized he was out of options and would have to contact his brother Paul. He hated relying on his brother, but the way things presently stood he could think of no alternative. Paul was currently at a remote site in the San Juan Mountains in southwestern Colorado, about to kick off another large right-of-way project. There was no way to reach him during the day by telephone, but Jacques finally did manage to contact Priscilla, his secretary, who manned 'Right of Way Inc.'s front office in Leadville. Priscilla served as the go-between when the brothers were out of town working separate jobs.

At these times, Priscilla's regular duties often took a back seat to mediation, because relaying messages between two men as volatile as Jacques and Paul was tantamount to being caught in a crossfire. Fortunately, Priscilla understood her role as go-between and handled the brothers with aplomb. She knew both had short fuses and she adapted accordingly. She was absolutely fearless with men and had a knack for toning down their heated rhetoric. She was also quite adept at massaging their male egos. In short, Priscilla was the slender thread that kept the brothers civil and speaking to one another. It was curious that even though she was vital to the smooth operation of Right of Way, Inc., the brothers remained strangely unaware of her key role and actual importance to their business. Priscilla was the unsung employee, the anonymous glue that kept everything together.

There were glitches, of course. After leaving Bowen Gulch, the previous afternoon, Jacques had attempted several times without success to raise Priscilla by phone. For some weeks, things had been slow in the office, and with nothing to do but sit by the telephone Priscilla sometimes took the afternoon off out of sheer boredom. Later that evening, he tried to reach her by CB, but again, without success. CB worked by line of sight, and was almost useless in the mountains during the daytime. At night, though, Jacques could often catch a skip off the ionosphere and reach Priscilla in Leadville. It was why the brothers had equipped her home with a CB unit. Priscilla hated the constant static, however. She claimed it drove her crazy. The woman probably had turned the volume down, again, too low to hear his call. This was no salve for Jacques' frayed nerves.

It was not unusual for the brothers to be incommunicado for days at a stretch. Jacques had no phone in his mobile office. Even while conducting the business in Denver, he was dependent on pay telephones. Earlier that morning, he had pumped nearly ten dollars in coinage into a public phone in the lobby of the state government building. Damned one-armed bandits.

We put a man on the moon but we can't even put a call through on the C-band. What the world needs is a goddamned portable telephone. Now why is that so difficult?

Once he made the decision to contact his brother, Jacques exited the interstate and finally reached Priscilla from a pay phone at a truck stop north of Denver. After explaining the situation he gave her emphatic instructions. "I need to talk with Hollinger too," he told her. "But Paul's the priority. Keep trying, all day, if need be. I absolutely must have a decision by tonight. We are down to the wire."

After finishing with Priscilla he pumped in more coins and phoned ahead to reserve a room at the Rustic Resort, which was famously located about half way up the south fork of the Poudre River on the road to Cameron Pass. Jacques had stayed at the quaint resort on numerous occasions, and thought well of it. At Rustic the raging Poudre slowed to riffles and meandered through a broad vale of fertile bottomland. It was a sweet spot, with large black cottonwoods along both banks.

His brother Paul was obstinate to the point of being pigheaded. Jacques had decided against trying to reach him directly. Any attempt at a conversation was likely to rekindle the same old issues over which the brothers so often feuded. Jacques was not eager to be on the receiving end of the dollop of grief Paul would certainly dish up when he learned about the eco-tage and the fine print in the Western-Pacific contract.

What a fucking mess!

His brother would probably explode at the news. Paul would need a few hours just to cool down enough to have a reasonably sane conversation. Jacques had been through it before, enough times to know better.

Best let Priscilla handle him.

The job in the San Juans was scheduled to start up any day. The site was a high alpine valley northeast of Durango, not far from the headwaters of the Rio Grande. The job would entail ground clearing for a new destination resort, including a major lodge and ski runs.

It was not unusual for the brothers to work separate jobs. The right-of-way business was lucrative, and there were obvious incentives to pursuing concurrent projects. To be sure, there were also risks. A company that allowed itself to become overextended could easily end up in a financial straitjacket. Jacques had seen it happen to a competitor and was determined to avoid similar missteps. Coordination was never simple or easy with two, or sometimes even three, concurrent projects. Often as not, the work sites were geographically remote. Managing projects at opposite ends of the state, sometimes at opposite ends of the region, could be a logistical nightmare. The brothers were often out of contact for days at a time; for which reason relatively minor problems could become serious in a hurry, especially when contracts involved deadlines.

Fortunately, the completion date for the ski resort in the San Juans allowed plenty of margin. If Priscilla could persuade Paul to delay ground clearing and come north with a skidder and some crew, they might yet avoid trouble with Ted Roe...

That surly son of a bitch...

The issue probably hinged on whether Paul had already committed to heavy equipment rentals in Durango. Heavy equipment was a big-ticket item, a major part of the budget for any large job. The St. Clair brothers, between them, did not own enough cats and skidders to cover multiple contracts, which is why rentals often were needed when working separate jobs. A signed rental agreement in Durango ruled out a postponement.

But good news was waiting when Jacques pulled in to the Rustic Resort, a message from Priscilla. The motel manager handed the note to Jacques when he checked in. It was brief and to the point:

PAUL AGREES. WILL COME NORTH ASAP WITH SKIDDER AND CREW. HUGS. PRISCILLA.

Finally, some good news. And about damn time.

Jacques set the alarm for 5:00 A.M., and turned out the lights. By the time his lids closed he was dead to the world.

THIRTY THREE

Fifteen minutes after the police car departed the campground, three shadows crossed the access road not far from where the deputy had parked. They tiptoed soundlessly through the night.

The trio had watched from concealment while the cop checked out the camp and called in his report. They were too far away to hear what the deputy said. It appeared that he was about to leave when for some reason he returned to one of the campers and went inside. Minutes later, he emerged with two men, one of them in handcuffs. After loading them into his patrol car, the cop finally left.

A boisterous party had been underway in the campground when Pinecone and his mates first arrived, about an hour before. No matter. The trio was in no hurry. They were prepared to wait it out as long as necessary, until the camp quieted down. The loggers had to sleep sooner or later. From the sound of things, the men had been drinking heavily. Everyone knew loggers were boozers. The wild party had actually been a stroke of good luck. The drunken fools would sleep that much more soundly, once they crashed.

As he waited, Pinecone glanced approvingly at his companions. Both were stout hearts.

Conditions this night were excellent. The moon would not rise for several more hours. The pitch dark was perfect. Tonight, they meant to finish the job they started two days before. Monkey-wrenching those skidders in Bowen Gulch had been a nice piece of work. No sweat. Putting the skidders out of action had bought time for Bowen Gulch, probably several days. Now, however, each hour was important. Without a working dozer there would be no road construction.

In recent weeks the Ancient Forest Rescue campaign had been steadily gaining support across the state. Things were moving in their direction. Polls showed the tide was turning. The previous day, two prominent state politicians had switched sides and were now publicly calling for the preservation of Bowen Gulch. Hope was rising too because Colorado Representative Pat Schroeder had introduced a new wilderness bill in the House. The bill included language that would cancel the proposed timber sale and add Bowen Gulch to the Never Summer Wilderness. Unfortunately, the legislation was stuck in committee, blocked by a senior conservative congressman known to be a strong backer of the timber industry. A whore for big timber. Yet, polls showed that if the bill ever made it to the floor of Congress it would pass by a wide margin. Why? Because the overwhelming majority of Americans supported wilderness legislation. For the moment, political posturing continued before the TV cameras, while serious horse-trading went on out of the spotlight.

Time was of the essence.

Pinecone's only regret was that he had not been there to see the look on that operator's face when he discovered the trashed skidders. That would have been almost worth the risk of arrest and prosecution. Mainstream environmentalists, of course, viewed this sort of indulgence as heresy. All of the big national groups, including the Sierra Club and Wilderness Society, had denounced "ecoterrorism." Indeed, they did so on a regular basis. It was nauseating how the nationals habitually kowtowed to the almighty dollar, that is, when they were not groveling before the god of private property. Two solid reasons why Pinecone and his mates had gone a different way.

As far as they were concerned, it was always open season on timber beasts. No quarter. Any operator willing to strip mine high elevation forest deserved whatever he got. It made perfect sense to hit them where they lived, in the pocketbook, yes, as hard and as often as possible. Make their business prohibitively expensive and maybe the bastards would abandon the sale, and think twice next time.

All the same, Pinecone knew that in addition to the obvious risks, sabotage remained a divisive issue, no mistake. Even among hard-core activists, opinion was sharply divided about monkey-wrenching, or eco-defense, as the Earth Firsters called it. There had been acrimonious debates, and little agreement. In the end the community of forest activists had been forced to agree to disagree. Pinecone was the first to acknowledge the important role played by the national and local groups. He had no problem working within the system. However, sometimes even the best system needs a gentle nudge, some elbow grease. To make an omelette you have to break eggs.

When the system fails, like now, you have to wing it. That's where we come in. What's the alternative?

The only reason he and his good fellows had not disabled the bulldozer on the previous occasion, along with the two skidders, was because the owner had installed a detachable steel jacket on both sides of the engine housing. The damn thing was padlocked on both sides. It had proved to be a simple but effective deterrent. They had not been able to gain access to the engine. Tonight, however, things would be different. He had come prepared. Pinecone had brought along a hacksaw and plenty of tungsten-alloy blades. They would cut the bolts that held the protective metal hood in place, then destroy the engine. Cutting the bolts would be tedious work. It might take them an hour, or more, but no sweat. There were three of them. They would take turns. He had also brought a small can of oil to lubricate the hacksaw blade and help muffle the sound.

The three now crept in closer. They were kneeling very quietly in the shrubbery at the edge of camp. For several minutes they waited in the pitch darkness, listening intently. The party had been winding down even before the cop's arrival.

Now, two loggers left the fire. Pinecone could not see them clearly. The men appeared to be moving slowly through the darkness toward the trailers. The fire had burned down.

"Listen," Mike whispered. "Do you hear?"

There were voices. "Yes," Steve whispered. Next, they heard a door close. Nary a sound. A blanket of silence enveloped the logging camp.

Steve put a finger over his lips. "Shhhhhh," then motioned for his comrades to follow him as he crawled forward. Slowly he inched ahead toward the dozer. Pinecone and Mike followed in single file. They wore black sweaters and stocking caps like commandos, and had even darkened their faces with black shoe polish. Not that they were fond of the war machine. On the contrary, their charge was a sacred trust, the defense of Nature.

When Steve reached the bulldozer, which was parked on a low-flatbed, he paused and waited for his companions to creep alongside. When they were in position, Pinecone produced a penlight and swept its tiny beam up and down the heavy equipment. He passed the light very briefly across the engine housing, then, clicked it off. Now that they were oriented, Pinecone led the way. He crawled up onto the flatbed trailer toward the business end of the cat. Very quietly, Pinecone raised his head up over the side of the D-6. Steve and Mike were right behind him.

Huh...?

A face stared at him out of the darkness. Pinecone was startled. Someone, it had to be a logger, was sitting in the driver's seat. The man was no more than a few feet away. Pinecone felt like he'd been kicked in the solar plexus.

Shit!

They were exposed! How did it happen? He fought back a wave of fear. Too shocked to flee into the night, he shone the penlight on Steve and Mike, and saw the same shock of discovery. Acting on impulse, he turned the light onto the face above him...

Whaa....?
The face wore a yellow Day-Glo mask.
Why does he look so familiar?
The man was smiling like a long lost friend.
"You boys here for the party?"

THIRTY FOUR

Camp was late rising next morning.

By and by, however, loggers began appearing in ones and twos outside Bobby's camper. No one had instructed them to do so. No order had been given. The assembly was entirely spontaneous. The men were still deep in vegetative euphoria. Indeed, the events of the previous evening were in progress. The loggers had a feisty look about them. Several were still wearing the Day-Glo. Most were half-dressed or in long underwear. Some were barefoot. A few were scratching their cojones. Hard-ons were general. Sourpuss Malone's pizzle was thrusting defiantly out of his pants, pointing skyward.

Two men showed up by virtue of the fact they never left. Shorty was still atop the boss's dozer, slumped over the controls, snoozing in the driver's seat. Credit the big man for never leaving his post. They found Fuzzy sleeping fitfully where he had passed out, his arms affectionately wrapped around a small aspen tree.

Tom had awakened to the fulsome sweetness of wildflowers, his nose immersed in the labia-like petals of Indian Paintbrush (*Castellja miniata*).

Wolfe and the Preacher were nowhere in evidence but their absence was not missed.

Bobby was the last to appear. He stepped down from his camper in regal fashion sporting a bathrobe and slippers, a jaunty red towel tucked in at his neck like an ascot.

There followed a collective improvisation.

None of the men had a clue about what had transpired (and indeed was still transpiring) but this hardly dampened their enthusiasm. Certainly there are no words for it, for how does one describe the be-mushroomed experience in human language? How do you articulate disembodied consciousness, or explain the extra-corporeal mind? One might as well try to square the circle, or capture a moonbeam, a rainbow, or a scintillating drop of morning dew.

Ineffability surely comes the closest, for it is just in the nature of the thing. "Just..." What a multifaceted sound. Can language truly do it just-ice? Or, are we "just" mouthing noises, platitudes, floundering about with our antiquated symbols and inadequate cliches? It was Heraclitus who, after hearing someone say that you cannot wade twice through the same stream, said, "No. You cannot do it even once!" Why not even once? Because that is just how it is.

Another philosopher, Ludvig Wittgenstein, expressed the same idea, though somewhat differently. Ludvig never ate mushrooms, as far as anyone knows; but he broke the same sod, whacked the same nail. Ludvig argued that existence is such a warped woof that only a very few things can be intelligently spoken of. His advice: become practitioners of Zazen. Try spinning the big wheel. Imagine a chimp silently pointing…

It was Charlie who finally assumed the role of group spokesperson, and vocalized what each man present wanted to know.

"So, Tom, what'd you put in that there spaghetti?"

But Tom was still basking in the mellow residuals. The best he could manage was a benign smile. The mystery would have remained but for Pissant, who spilled the secret, "Don't you b-b-boyz know?" he stuttered. "It was them m-m-muchrooms."

Silence.

"You mean, in the woods?" Charlie finally said.

Pissant nodded.

"Well, fuck a duck!"

After considerable commotion, Pissant produced a bag of the said m-m-muchrooms. Charlie wanted a closer look and stuck in his paw. Sourpuss was there too, looking over his shoulder. Charlie retrieved a small fungus, held it up rather daintily in his big rough hands, studied it a moment, then, quickly passed it over to Sourpuss, who hesitated before taking it. The logger oh so gingerly lifted it to his nose and gave it a cautious sniff. Shuddering, he hurriedly passed it on to Dipstick. "Here, you take it," he said, wiping his hand on his shirt. The daintiness caused sniggering all around.

"Don't give me that thingee!" Dip said. "Oh gross."

"Thingee!" a voice said. More sniggers.

The loggers were not yet convinced, not entirely. Dipstick handled the cap in much the same manner, which is to say, cautiously, before he too passed it along. Each of the others did likewise, sniffed and studied the thing, then, hurried it on to the next fellow. Each seemed visibly relieved to be rid of it. So it went, until the host came back, full circle, to Charlie who was still holding the bag.

"What the hell?" he said, and promptly ate it.

"Even if it kills you, right, Charlie?"

"Damn straight."

Charlie was still chewing to a buzz of nervous laughter. The loggers quietly watched as he gulped it down. Charlie made a face, and they looked relieved when he did not drop dead on the spot.

"How's the taste?"

"Ugh!" he said with a loud belch.

"They ain't 'thingees,'" Pissant said. "They's the food of the gods. M-m-manna from heaven."

This touched a nerve. It was a Eureka moment. Their faces lit up at almost the same instant. They were now moving together up the same rapid learning curve, riding the same resurgent wave.

It was a kind of group transmission, like the hundredth monkey.

"Don't hog the bag," Jimmy said and grabbed it away. The big man produced a cap and held it under his nose. "What did you call 'em? Food of what...?"

"G-g-gods, Fuzz. Food of the g-gods."

"How many we got in there?"

Jimmy made the count. "One, two, three...four left is all...That's it."

"Damn! Not enough!" said Dipstick. Now he handled a mushroom with the air of a trained mycologist. "Not nearly enough. We'll have to round up more..."

"Yep. A lot more."

Four different loggers quickly downed the remaining caps. Even in the act there was loud whoop-dee-doo, a kind of group affirmation that sealed the collective insanity.

They had arrived.

What happened next was foregone, what one might expect of a group of uniquely different individuals, all of whom awaken one morning with the same thought uppermost in mind, all having dreamed the same dream. No surprise that they were tracking in the same groove. Had they not already been on a converging trajectory for many hours?

There would be no start-up at Bowen Gulch this fine morning (nor the next, nor the one after that). It was just not going to happen. No, the firm grip of gainful employment had been irrevocably broken. The reality principle had lost its power over the crew. The concept of work had been dethroned. The verdict had come in. Their collective life sentence (the forty-hour workweek) had been commuted by general acclamation. Work and play now resembled the phases of a gestalt. To the extent that one (play) waxed, the other (work) must wane.

To the group mind all of this was self-evident.

There was also a dawning sense that the good times had arrived. The same proverb like unspoken fire was dancing on the tip of every tongue. But the actual ratification, when the just-described subterranean process of communal "group think" finally broke through and became fully conscious in a collective sense, did not happen until Charlie, again, gave it expression.

"If that ain't some ass-kicking shit I don't know what!"

"Amen," echoed Dipstick.

"It's happening, boys!"

"Fungus-amongus!"

"You got that son of a bitch right!"

There was a loud "Wooooooopeeee!"

And that is how the group decision occurred to blow off work and get down to the important business of hunting up more of the m-m-muchrooms.

Remember, these were not doped out beatnik poets in dark sunglasses and sandals.

These were not bohemian artists out on the raggedy edge.

These were not the sons of the literati flaunting their avant-garde lifestyle.

These were not zonked out acid-heads tootling pedestrians from some psychedelic bus.

Nor were these Dead Heads panhandling admission to the next rock and roll orgasm in the sky.

No, no, no, no, no, no, and no again. No, these were loggers. L-O-G-G-E-R-S. And if such a thing could happen to such staid individuals, well, could it not happen to anybody? If it could happen here, in the bosom of the beast, in the veritable heartland of rip-it-up America, might it not happen anywhere? Which is to say, mothers and fathers, if you know what's good for your precious babes, you better manacle them. Lock them up.

Because no place is safe from the groove machine.

So it went, all manner of spontaneous camaraderie, laughter and joking around, horseplay, hoots and howls, arm twisting, shoulder slapping, back thumping, tussling, and so on.

That is, until Bobby's girlfriend Rusty poked her dusky blonde head out the door of Bobby's camper. None of the men had seen her up to this point. It was Rusty's first appearance in camp. To say that her lovely countenance caused a stir would be a considerable understatement.

There was a stoned silence.

The sight of her was like a seed crystal dropping into a super-cooled fluid. That is what started them moving, the stone that gathers no moss.

As if drawn by invisible threads, five of the loggers, including Tom, hurriedly crowded into Thurston's Jeep Wagoneer.

No one had verbalized any collective intent. No need. The group mind was leap-frogging ahead. The men (each one as horny as a billy goat) already knew the destination. They were headed for the Kawuneechee Lodge at the foot of the valley, some six miles away, on the road to Granby, the site of the nearest public telephone. The purpose of the sudden expedition: to summon wives, lovers, and girl friends from Granby, Fort Collins, Loveland, Kremmling and points beyond.

Thurston and company did not get away clean. By the time Jimmy fired up his rig and backed out into the road, Pissant and Shorty were swarming against the glass and pounding on the hood, with urgent appeals to "Hey while you're at it, Dip, how about asking Sally to bring her girlfriends, and her kid sister too, if she has one, and, for that matter, anything in a dress (aunts, nieces, mothers, no problem, even grandmothers). Oh, and don't forget to pick up a few cases of beer. Supplies are running low."

When the messages had been duly conveyed, the Wagoneer disappeared down the road out of camp en route to the Kawuneechee.

Sourpuss and Kermit jumped in their own rigs and followed solo.

Shorty and Pissant were left standing in the road amidst their own thoughts, the dust of departure settling upon them.

But not for long.

Soon, these left behinds returned to camp, pulled on their boots, pants too, and armed themselves, each with a plastic garbage bag. By unanimous assent of the group brain, Shorty and Pissant had become co-chairmen of the m-m-muchroom collection committee. Pissant was the mushroom expert. He knew where to find them.

Presently, the two herded up and splashed across the little creek behind the camp into the deep timber to gather up Nature's bountiful harvest.

During all of this feverish activity, Bobby Lighthorse had hardly moved from his spot where, still in his bathrobe, he had watched the wacky business. Mushroom mania had carried the day, sweeping everything before it.

Things were not as they appeared, however. Bobby had not been left behind. On the contrary, he was, even now, the scout leading the pack, blazing trail for the others.

When the departures and rushing about hither and yon were done, he strolled toward his camper. Bobby was already feeling the first effects of the half-dozen mushrooms he had consumed earlier that morning in lieu of solid fare. Three times as many as anyone had downed the previous day. Rusty was there to greet him, smiling sweetly as she poked her cute little head out of the camper door. Bobby could not see the rest of her, but he suspected she was still in her nightie. Visions of sugarplums levitated behind his eyeballs.

Ah, she is a lovely thing, more lovely this morning, I believe, than I've ever seen her.

"Golly, what happened?" Rusty said in her pert manner. "It got real quiet."

"Yes." Bobby said, looking up into her attractive face.

"Where's the boys?"

"Gone this way and that, honeybun." He motioned with two hands in opposite directions. "But don't fret, they'll be back soon enough. Let's enjoy the peace and quiet while it lasts."

Bobby was wearing thin-soled slippers, however, and as he stepped toward the trailer he came down on a sharp little stone. "OW! OW! OW! OW!" Grimacing, he hopped about on one foot and nearly fell over. He pulled off the slipper and rubbed his heel, chaffing at the smart.

Rusty giggled, hand over her mouth. "Well I declare, you logger men aren't half as tough as you let on." She had a fresh tongue, Rusty did. She was a southerner, Georgia born. A peach.

"That's my girl," Bobby said as he stepped up into the trailer. When she turned he gave her a playful slap on the behind.

THIRTY FIVE

Jacques St. Clair streaked by the Kawuneechee Lodge at seventy miles an hour. He was slowly gaining elevation as he moved up the valley. As a general rule, the boss drove with his foot on the floor; however, on the return from Rustic he had stayed under the limit most of the way. He could not afford to lose the lowbed driver who was following him with the rented skidder. Due to the heavy load the guy could barely keep up as it was. However, they were now within a few miles of Bowen Gulch and the boss's foot got the better of him. The lowbed driver could not go too far wrong. Jacques had been given him directions and the man would catch up. The boss stomped on it.

He had awakened in a nasty temper and was now in a general funk, for during the long drive a noose of dread had been tightening around his thoughts. The pisser was knowing that but for the incompetence of others his logging operation would be on track at this very moment, yes, running smoothly, humming right along. But who can foresee such things? Who can predict the untoward actions of eco-nuts and arrogant attorneys and high-handed bureaucrats that so easily derail the best-laid plans of men? No wonder things fall apart.

Jacques knew he'd been away too long. It had been less than two days but that was still too long to leave a project unsupervised. As he knew only too well, a timber operation has a way of going to seed when the boss is away. He had seen it happen before, yes, too many times, and he was not about to feel complacent now, even with a stand-in as able as Francis Delacour minding the store. Under ordinary circumstances Francis could be counted on to keep the lid on a job. But the project at Bowen Gulch had been anything but ordinary.

It's been a circus.

Jacques began to anticipate what might lie beyond the next blind curve...

Where will the next blow fall? If something can go wrong it will.

Thus saith Murphy's Law. Or was it the Peter Principle? For the life of him Jacques could not remember.

Is it not human nature to weigh the prospects for success against the likelihood of failure? What is life but a parade of screwed up situations crying out for redress or resolution? A sage and honest man, accurately assessing the fleeting nature of human existence, will consciously cultivate a healthy sense of detachment, disavowing worldly success and failure alike. But alas, few men are wise, least of all when faced with life's slings and arrows. Most men will brood in the face of adversity for the simple reason that most men hope for, no, a stronger word, most *insist* upon a happy outcome, deliverance from the mulish face of outrageous fortune; and to this human rule Jacques St. Clair was no exception.

Fresh doubts and worries now assailed him, crowding out the last vestiges of his peace of mind. As the boss's thoughts hit bottom he suddenly hated the work. The hassles were just not worth it.

Fuck it all! Who needs it?

For a moment Jacques visualized just walking away. Why not liquidate? He could do it. Sell his share of the business to his brother. The tantalizing prospect of freedom now danced before him like a voluptuous woman that a man is free to desire but can never possess. Mesmerized, Jacques stared straight ahead past the tip of his nose into blank space even as he drove, wallowing in fantasy like a boar driven to the edge of insanity by a sty full of sows in perpetual heat.

Whom did he encounter peering into this darkest rummy corner of his soul? Why, his better half...

Ahhhhh sweet Sarah.

His doting wife was loveliness itself. He and Sarah were now arm-in-arm, standing together, watching their six-year old son Isaac ride his two-wheeler. The boy was the apple of his eye, the pride of his life. Only last month he had removed the trainer wheels from Isaac's bicycle. Now, as he watched his son in his mind's eye Jacques wanted to sweep the lad up in his big rough arms.

In that instant the boss was back, fully in command, scoffing at himself.

Screw the tree huggers and the fine print and the escape clause and the parasitic lawyers that never produce a thing in this world except trouble for honest working people. Screw them all!

As Jacques rebounded from his dark reverie he began to contemplate his options from a more positive angle. Damage control was still possible. He might yet avoid a financial bath. Only a fool judges it too late to cut his losses. A man likes a positive spin. Is not trouble just another word for opportunity? Does not each problem contain within it the germ of a solution? Yes, it comes down to how a man choses to see it. One can decide to view the daily cup as "half full" or "half empty." Take your pick.

This was not the first time Jacques had faced trouble on a job. He had been pressed hard before, a time or two. Once, he had been backed to the wall.

Yes, there was that job in Wyoming.

It was a big right-of-way job for a power line through the Wind Rivers, when the weather went to hell in mid November and the diesel turned to jelly in the tanks. That had been one cold bitch in hell. The project was also similar to Bowen Gulch in that it had been complicated by fine print.

Double-digit bullshit! But I bested the bastards!

Yes, Jacques had waltzed to the bank with the largest net take ever for Right of Way Inc. Even his over-cautious brother had been impressed. Didn't it always come down to last resorts, the final inch? Was this not the true measure of a man? His mettle?

Jacques gave it more gas as he rubbed his abdomen. He had left Rustic in such haste that he skipped breakfast, a mistake as he well knew. The boss's insides were always prone to enzymatic self-destruction. His nervous stomach was trip-wired to his mercurial moods, and now he was feeling the familiar ferment, accompanied by those tumbling belly rolls and somersaults that sometimes made him feel old and gray before his time.

With his left hand on the wheel the boss reached across the dash and snapped open the glove compartment. Brushing aside a tape measure and a flashlight, he groped under some maps until he found the Tums. With the bottle in his free hand he pried off the cap with his teeth and popped several pastel tablets into his mouth, he didn't bother to count, as he hit the gas, veered across the median stripe into the opposite lane, and roared past what looked like a family in a blue station wagon with Iowa plates. They were creeping along at a lame fifty mph. As Jacques cruised back over the line, chewing antacids, he looked in the rear view shaking his head.

Tourists...

For the next few miles there was no traffic, just the way he liked it, the open road. But, now, he passed a vehicle going the other way that looked familiar, and it was followed by two others.

Jacques caught a glimpse of the lead driver out of the corner of an eye, only a glimpse but that was enough. Maybe it was the wise-ass grin on the man's face, or something about the mug.

What...!

Staring into the rear view he missed the next curve and careened across the stripe into the other lane and nearly ditched his pickup. Fighting for control, Jacques swerved at the last instant. That kept him on the shoulder, but even so, he nearly lost it. With a *screeeeeeeech* the truck jumped back onto the highway tracking rubber and fishtailed down the road. Fortunately, Jacques was an able driver, and luckily there was no oncoming traffic. He swerved until the brakes finally took hold and rolled to a stop. Through it all one eye had been glued to the rear view.

I know that rig. That not only looks like! Son of a bitch if it isn't Jimmy Thurston in his Wagoneer going the other way like a bat out of hell. An' it looks like...yes, he's got...what? three...four of my loggers with him. Plus two other rigs..

Jacques found all of this difficult to assimilate.

That's more than half of my crew heading down the valley at ten in the morning, away from the job site...where they ought to be dropping trees.

He checked his watch, then, with his sleeve wiped away the sweat on his forehead. For good measure he popped two more Tums. His stomach was doing those freaking half-gainers again.

Fifteen minutes later he reached the field office at the main landing at Bowen Gulch. Braking hard, he spewed gravel and slid to a stop within ten feet of the trailer. Popping the stick out of gear, he yanked on the safety brake and without turning off the engine slid out the door and headed for the office, calling his foreman. "Francis!"

No reply. There was no one around. Jacques poked his head inside the office. The place was forebodingly empty.

Where can that man be?

He fought back a mounting rage. His chest began to palpitate in asynchronous counter rhythm to the jazzy back-beat under his belt. Suddenly dizzy, Jacques leaned against the trailer to steady himself as a blinding flash of white light moved up his aorta and carotid artery, stabbing into his head. The shock passed, however, and he sighed with relief as he saw Francis Delacour coming out of the timber on the far side of the deck.

"Francis!"

Jacques was instantly annoyed by his Foreman's casual demeanor. The man appeared *too* calm.

"Morning, boss. Did you get the skidders?" The man's pleasant tone and manner grated on the boss's frayed nerves.

Does he never react? Francis would sound upbeat after being mugged out of his skivvies!

"I could only locate one skidder," Jacques said, spitting out the words. "The driver's right behind me. But *that* can wait. What's going on, here? Five minutes ago I passed Jimmy Thurston and half of my crew going *down* the valley, headed for Granby."

Francis stared back not comprehending, his face a blank.

Now came another shock. Reeds were parting inside Jacques' head. Slowly he was becoming aware of what his eyes and ears had been trying to tell him, indeed, had been shouting at him ever since he climbed out of his pickup. Jacques looked around.

Where are the others?

Yes, that's it. Wake up to the fact there are no other vehicles in sight. No rigs meant no crew, no business as usual, no start-up. Where are the rest of the men? The jimmied skidders were still parked at the edge of the landing but otherwise the deck was deserted.

Jacques paused to listen.

Yes, wake up to that, too. The problem is that the woods are strangely silent, too silent, absent the noisy music the boss has learned to love so well. The healthy din of screaming chainsaws telling him a crew of men are on the job, getting after it.

"The woods," Jacques mumbled. Yes, Jacques, the woods! The timber! Where is the sonorous medley of men at work that is music to your ears? The sweet symphony of loggers tapping out their screaming Stihls, Huskies, Partners, Jonsereds, and Homelites to the max. The lovely clamor of men dropping trees and limbing branches. Oh those beautiful discordant harmonies! Timber on its way to becoming logs, soon to be loaded up and sent down the road. The sound of money, Jacques, greenbacks, coins clinking, silver rattling and rolling, hard-earned interest accruing on principal, liquid assets piling up like wood chips. The sweet sound, God willing, that will put you over the top and make you a wealthy man, or, at any rate, rich enough to retire before the age of fifty. Thereafter to kick back and gather in the good life. Recreate? Travel? Hell yes, whenever and wherever you like, maybe to winter in sunny Arizona or southern California, or even fly down to Oaxaca if you have a mind to, and take the wife and kid. And summer? Yes, summer in France and Switzerland, visit the Bernese Alps and the great cathedrals, and Paris too, the Louvre, and, while you're at it, track down the family's ancestral village in Normandy. Or maybe just fly fish the high lakes in the Sawatch range if you take a notion. Or hell, for that matter, just stay home and watch football, or do nothing – laze around the house all day in your PJs. Is this not the American dream? Is this not what the labor and sweat is all about? Why we strive eight to ten hours a day, five or six days a week, 50 weeks, year after year? Is it not why we put up with the hassles and the bureaucrats, the slings and the arrows? Yes, of course it is. This is why it matters greatly that the woods are as quiet as a cemetery, this morning, and why where there ought to be screaming saws, there is nothing...

"Boss, I..."

Jacques waved Francis silent. "Shhh, wait!" he said. Side-by-side the two men stood, Francis the ramrod biting his lip and Jacques the boss champing at the bit, his ears straining toward what he thought and hoped would be there, what *must* be there.

Maybe they used the other Forest Service access road. Maybe they parked higher up...

"Do you hear?"

Muffled by distance came the faint but unmistakable whine of a chainsaw echoing up the valley. The sound was feather-light, soft as velvet on the morning air. The saw was a long way off. But no matter. One saw.

Jacques waited, still counting. But that was all. That was it. No more. Only one saw. ONLY ONE! One saw where there ought to be a medley. One saw a symphony does not make.

As if reading the boss's thoughts, Francis said, "That's not us, boss. It's just some guy cutting firewood. I already checked it out. The crew didn't come in on the other road either. No one showed up for work."

The words hit Jacques in the gut. He grabbed Francis roughly by the shirt. "Where's the rest of my crew?" he shouted in his face.

"Hey! Whoa there, boss," Francis said gently, matter-of-factly, in the lowest key he could muster. Francis was concerned by this rising ire in the chief's voice. The tone was as dangerous as a chisel tooth. "Simmer down, chief. No cause to get riled."

Jacques loosened his grip. He knew his foreman was right. Yes, right as ever. Francis was always right. Thank heaven for cool-headed ramrods like Francis Delacour. Struggling to regain his composure, he brushed his foreman's shirt with his hand as if to erase the outburst. He took several deep breaths. The heat drained from his face. Calm returned. Jacques motioned down valley. "I just passed half the crew," he said. "Would you believe they were headed for town?"

"I know, boss.

"You do?"

"You just told me."

"I did?"

"Look, chief," Francis said, "I was just this minute on my way over to camp, to find out what's going on."

"Huh?"

"When you pulled in." Francis paused, then continued. "Boss, yesterday evening, Rosemary went into labor, so I had to rush her to the hospital. But when I left camp everything was quiet. No problems."

It took another moment for the words to penetrate the clouds. Finally, Jacques asked with genuine concern, "How's Rose?"

"Fine, boss. Just fine," Francis was clucking like a proud rooster. "We have an eight pound girl. She arrived at five this morning. I'm a dad." Francis was beaming from ear to ear.

"Congratulations," Jacques said. He was smiling too. "That's great news." He gave his foreman a friendly punch. Momentarily reveling in the good news of fatherdom, the two men embraced in a warm hug. However, the good feeling was short-lived. "Didn't the men know today is start up? Hell they should have, I told them a dozen times."

"I did too, boss."

"So, where the fuck are they?"

Francis shook his head.

Jacques looked at his watch. "They should have been here over an hour ago..."

"It's weird, boss. I don't know *what's* going on. Something."

"Damn," said Jacques. "You think they're down with ptomaine?"

"Don't know. They..."

"What?"

Francis told Jacques about the previous evening's all-you-can-eat spaghetti feed. It was enough to send the boss's dark thoughts galloping ahead. If Jimmy and crew were down sick that would explain why they skipped work. It might also explain why they were headed for town. Maybe the men were on their way to the medical clinic in Granby.

Good Christ, just what I need, my entire crew out of action with stomach flu, or worse.

Jacques had seen it happen, yes, more than once, seen the biggest toughest brawniest loggers laid up for days, weeks even, racked by dysentery, helpless as babes, flat on their backsides, that is, when they were not beating a path to the can. Stomach flu could put an entire crew out of action. Not a pleasant thought.

Murphy's Law again?

The risk of food poisoning was ever-present with men who are not in the habit of bathing regularly or cleaning up after themselves. It was why Jacques always made a point of reminding the crew to wash with lots of hot water after eating. "And dammit! Use dish soap." Not that the bastards usually listened. So much wasted breath.

"Only one way to find out," Jacques mumbled.

"I was casing the lower unit," Francis said. "Just now. I wanted to make sure they'd finished the flagging. Looks like everything's ready." He paused. "Maybe I should have checked the camp, sooner."

"We'll do it now."

Francis nodded. The two men jumped into the boss's still purring rig. Jacques released the hand brake, threw it into reverse and backed out sending gravel flying. He jammed the stick hard into second, forget first, and gunned the engine. As he picked up speed he shoved it into third, grinding gears. They made all possible speed down the Forest Service gravel road toward the logging camp trailing rocks, exhaust, and dust, not to mention the acrid smell of burning clutch.

THIRTY SIX

The logging camp was serenity itself. From somewhere back in the woods came the drilling of a downy woodpecker. The sound had a dreamy quality like a stroke of balm on the inner ear. The aspens and firs rustled ever so gently, caressed by the softest of summer breezes. More balm; the sound of solace in a world gone mad, spinning off its rocker.

In Bobby's camper the peaceful air stirred the half-open window shades. However, into this sea of tranquility was about to fall a red-hot poker, seething and hissing, violent and intrusive as a squeaky wheel.

Rusty was lying on the bed beside her man feeling the much-anticipated first effects herself of *los ninos fabulosos* when from afar came the distant low roar of a motor. From half a mile away the sound had no substance. It was hardly more than the suggestion of noise. Whether her man also heard it was doubtful. Bobby was no longer aware of sound, as such. In his case bed, trailer and world had already merged into an ego-less avalanche of liquid plasma.

Presently, the sound of the motor grew louder. Rusty propped herself on an elbow and now distinctly heard the metallic crunching and grinding that even she recognized as the hurried and reckless downshifting of gears. This was followed by the screeching of brakes and tires on gravel. No doubt about *that* sound. A vehicle, probably a truck, had just arrived at the logging camp. Now came the racket of truck doors opening and slamming shut, and the loud noises that men make when they're in a big stew about something.

"Bobby, somebody's coming," Rusty said. "And whoever they are they're in a hurry." Bobby half-opened one lazy eye but promptly closed it again. He rolled over and buried his head under a pillow, Bobby's way of saying "no" to external reality.

Now came the thunderous sound of the foreman's deep bass voice booming through camp; and above it the no-less-insistent bark of command, the voice of Jacques St. Clair, crew boss.

"Bobby," Rusty said, shaking his shoulder. However, the gruff commotion and even Rusty's gentle hand on his shoulder was like files screeching on his heightened senses.

For another short eternity he lay in peaceful stillness, absorbed by the inner symphony of the spheres, much too content to be bothered by external trivia. Rusty shook him again, harder this time.

"Bobby! Come on! You got to get up."

"Uhhh?"

At last he stirred. Moving in slow motion, Bobby pulled himself up and sat on the edge of the bed. He rubbed his face and eye sockets. Slowly Bobby gathered himself and began the Herculean chore of dressing. He reached for his robe, but tossed it aside.

"No, not that," he said. "Seen my pants?"

Rusty handed him his trousers. Without bothering to pull on his underwear, Bobby rose from the bed. He did not even attempt to put on his shoes and socks.

"Oh bother."

"My goodness! Are you large!" Rusty said, giggling at her man's tumescence as he shuffled toward the trailer door.

"I'm here, Jacques!" Bobby attempted to shout as he shoved a bare foot into the leg of jeans. "Over here!" However, the words came out in a barely audible whisper. To Bobby, the croaking sound of his own voice was like the far-off sound a pebble makes when you drop it into a deep well, nothing, nothing, nothing, then out of the void, *ker-plunk.* To Bobby it seemed as if this remote *ker-plop* was the fathomless sound of time itself. Nonetheless, he was determined to make himself heard from the depths of this deep cave. Marshaling himself against the sweet chords and the melodious strains inside his head, Bobby poked his head out the trailer door and called again. This time he succeeded in projecting his voice with force.

"Over here boss!"

It was only when he got to the zipper that he understood what Rusty meant.

THIRTY SEVEN

Jacques St. Clair and Francis Delacour strode impatiently through the logging camp calling out names and peeking into trailer doors and windows. The place was strangely deserted. There was no one around, that is, apart from Dipstick's mangy blue tick coon-hound whining and groveling at the boss's feet.

"Get out of here!" Jacques growled. "Out of my sight!" The dog slunk away, tail between its legs. "That shit-eating excuse for a dog would be better off put out of his misery."

Francis was concerned about the boss's rising blood. "What's this?" he said, staring at a large red bull's eye painted on the side of Dipstick's camper. He stared at it in disbelief, then noticed other scribblings. One entire side of the trailer was covered with graffiti. "Boss, look at this."

Jacques was three steps ahead of him, but turned back. "What now?" He stared at the graffiti. "What in the fucking hell?" The two men gave each other a quizzical look. Suddenly, they heard a voice. Someone calling.

"Boss, you hear *that?*"

"Yeah. Where's it coming from?" Between them, Jacques and Francis could not be sure. No man in the woods, not the sharpest-eared hunter, not even Hawkeye, can accurately cipher the direction of a sound or a shot if it comes only once. Stock still they stood, their ears pricked, waiting for a second report. After a moment, it came.

"There!" They had their fix. The men changed direction, Francis now in the lead. "This way."

The voice had come from the far side of camp, and that is where they found their quarry, Bobby Lighthorse. The man was standing outside his camper, barefoot and bare-chested. He was attempting to zip up his blue jeans, but was fumbling and having a time with it. Four eyes dropped and perfunctorily took in the lump in his britches, then, moved on.

Jacques noticed the graffiti scrawled on his dozer. For a moment he stared at it, as if not believing his eyes. He turned to Bobby, fuming.

"Bobby, what's this? Who did it?" he said, motioning to the cat. "Eco-jerks?

"No, boss."

"Who, then? And where's my crew?"

"Gone."

"I can see that. But ... where?"

Francis was standing well back. He was worried. The boss was near the boiling point.

Oh shit...

Silence from Bobby.

"This *is* a work day," the boss said, "or have you forgotten?" Suddenly sensing that this stupid sparring was getting him nowhere, Jacques tried a different tack. "Son, let me put it *this* way. What are *you* doing here in camp, when you ought to be dropping trees?"

"I was in the sack..."

"In the *sack?* Are you sick? You got the runs?"

"No, boss."

Smoke was almost visibly coming out of Jacques' ears. To side step the enormous rage mounting within him, the boss glanced at his wristwatch and frowned. Now, the timbre of his voice changed. "It's after ten-thirty," he said through his teeth. He recalled the lump. "You got a woman in there?"

No reply. Silence. The moment hung by a thread.

Francis knew that when the boss got this way, things were far along, very far along, indeed. But although he knew an explosion was coming, still he hoped it might be averted. One had to admire the way Jacques wrestled with himself to corral his temper, the way he often did manage to contain it. The very fact that he expended such effort was admirable.

"So what were you doing in bed?"

After a silence that seemed too long, Bobby said, "You sure you want to know?"

"HELL YES I WANT TO KNOW!" the boss shouted back. *"T'ES COMPLETEMENT DINGUE?"*

Francis winced. He winced again when the kid yawned. It was one of those long drawn-out lazy yawns, the kind that cats make, slow and relaxed, deep and unfazed, self-indulgent and impertinent. Cats! The one animal the human race had domesticated without ever taming. The inscrutable bastards.

That's all she wrote...

Francis took two steps back to avoid the shear zone. No point standing in the path of destruction.

"DON'T BACK TALK ME, BOY!" the boss screamed, losing it now. Jacques ripped off his hat and flung it on the ground. "YOU SMART ALEC KID! I'LL TEACH YOU SOME RESPECT! WHADDYA THINK THIS IS, A BOY SCOUT CAMP? I GOT A SCHEDULE AND DEADLINES TO MEET HERE! THERE'S WORK TO BE DONE...W-O-R-K! OR DON'T YOU KNOW THE MEANING OF THE WORD? I THOUGHT I HAD ME A CREW OF LOGGERS UP HERE BUT GOD DAMMIT, I GUESS I WAS WRONG. WHAT I GOT IS A BUNCH OF JERK OFFS. LIGHTHORSE, I GOT A MIND TO FIRE YOUR WORTHLESS ASS! WHADDYA SAY TO THAT, BOY? HUH? BY GOD, I THINK I'LL DO IT! GOT NO TIME FOR LAGGARDS AND WUSSIES!" Jacques pointed his finger like a pistol. *"FOUS-MOI LE CAMP!* AND I WANT YOUR LAZY ASS GONE BY SUN UP, TOMORROW MORNING! GOT THAT? SO YOU MIGHT AS WELL START PACKING YOUR GEAR!"

And that was that. The boss turned and stormed off, a slipstream of superheated ions sparking the air behind him.

As the boss went by Francis reached out a hand and opened his mouth to do or say something helpful. Thinking better of it, he stepped aside and made room for the passing tornado. For another moment Francis stood flat-footed scratching his head, staring at yellow and brown aspen leaves on the ground at his feet, wondering why ("For heaven's sake...") they reminded him of his wife. On any number of occasions, he had seen Jacques lose his temper in just this way. Yes, and on a few of those occasions he had seen the boss angrily throw his hat on the ground. However, he had never until this day seen the boss too angry to retrieve it. Francis stepped forward and did his duty. Leaning down, he picked up the hat and dusted it off. Now, staring at Bobby, Francis opened his mouth again to say something helpful, anything to ease the tension; but the moment was past redemption. With a shrug, he turned and followed the chief back to the pickup.

By the time Jacques reached the truck he was already having second thoughts and misgivings about losing his temper. He didn't have to think to know he'd made a mistake blowing up at Bobby like that. From long experience Jacques knew that crews do get out of line. On occasion. After all, boys will be boys, but once in a while, a boss had to take decisive action to put things right. Jacques did not enjoy being the heavy in these occasional extreme situations. If the truth were told, he was the peaceable sort. He disliked confrontation, finding it distasteful. But damned if he would back down from trouble either. Usually the problem centered on one bad apple spoiling the barrel. Making an example of one man usually was enough to turn the herd back on course. Excise the bad egg and you generally had the problem whipped at the root. Nine times out of ten.

What had Jacques stumped now, though, was not simply the fact that he had failed to learn the whereabouts of his crew. He still had no clue what the problem *was*, let alone which bad seed was behind it. Nor did he know who had defaced his cat and half the campers with Day-Glo paint, or why.

His gut told him that Bobby was not the lone culprit. For which reason singling out the boy and firing him had not only been foolish, it was just plain dumb, about as pointless as stepping out into a pitch dark night and taking an aimless potshot with his thirty-thirty.

Hells-bells, what's a man supposed to do? Just sit there and take it?

Such were his disordered thoughts as Francis slid into the cab beside him, plopped the cap on the seat, then waited in silence for the boss to speak. It was always best to let Jacques get whatever was eating him off his chest, in his own time.

Thank heaven for loyal, soft-spoken, and long-suffering men like Francis Delacour. The chief turned to his foreman without consciously noticing the hat, regret and frustration written all over his face.

"Would somebody be so kind as to tell me what in the Sam Hill is going on, around here? Is something happening, or am I going out of my gourd?"

"You got me, boss."

"I don't know about you," said Jacques, "but until this morning, I thought I'd seen everything."

"We got ourselves a situation, alright," Francis said. "But *what* ... I dunno. I'm in the dark as deep as you."

Jacques started the truck and headed for the barn. However, before he had gone a quarter mile he met Jimmy Thurston and company coming up the access road toward the camp. Apparently the boys were returning from Granby, or wherever...

"OK. Good," Jacques said. "Now we'll get some answers." The boss braked and stopped the truck in the middle of the road, as if concerned that Jimmy might fail to see him and drive on by. To make extra certain, Jacques leaned out and waved down the Wagoneer.

What he saw was anything but delightful.

Jimmy was at the wheel, alright. The boss also recognized the familiar mugs of four other loggers: Charlie McCoy, Dipshit Dugan (or was it Dipstick?), plus Red Callahan and Tom Lacey. Two of the men had Day-Glo on their faces...

What in the hell...?

Not only were they drinking beer at ten in the morning, on a workday, it was evident they had already put away a few. Incredibly, the men were flaunting it, as if they were on holiday. To say that Jacques was displeased by all of this would be like saying Poseidon was slightly peeved with Odysseus. He was seething; but he would sooner freeze in hell than lose his cool three times the same morning.

"Hey boss, how's it going?" Dugan said for openers. The guy was smiling in that infuriating way of his. Thurston and Red and the others were the same, grinning like there was nothing out of the usual. Jacques started to ask about the Day-Glo but thought better of it. Best not go there. The noxious whiff of mutiny was in the air. Jacques could smell that vibe anywhere. It was unmistakable.

"Small world it seems," the boss said. The smile on his face was thin as a wire. He held his tongue on a tight leash.

"It sure is, chief," Thurston said. "We thought maybe it was you back there we passed, above the Kawuneechee."

"None other, Jimmy."

"Yeah, we thought it was."

What was puzzling the boss about this menagerie of faces, were the grins that kept coming. "Party time, boys?"

Thurston stared at the boss. Something about Jacques was different, this morning. All of a sudden Jimmy knew what it was. The chief was not wearing his hat, the one with the CAT logo. Strange. Jimmy could not recall ever seeing the boss bareheaded. For a moment he studied Jacques' black hair, flecked with gray, noting the way the boss parted it on the right side, instead of the left. The dark hole in the middle of the boss's face seemed larger and more conspicuous than he had ever seen it. "Would you like a cool one, chief?" Jimmy said good-naturedly. "We got plenty." Thurston turned to the men in back. "Hey Dugan, pass the chief a Bud. One of them cold ones."

"No thank you, Jimmy," Jacques said curtly. "I'm afraid you misunderstood me. What I meant to say was, isn't it a *little early* in the day to be drinking?" His tone was a smidgeon on the sharp side, and brought no response from the Wagoneer, only silence, that is, until someone in the back sniggered, a spark that set off the whole caboodle.

The hysterical whoop was like a rude fist in Jacques' face.

That woolly headed Dipshit!

"Did I say something humorous?" Jacques sallied back, his tone as unyielding as granite. "I'd really appreciate it if one of you boys would *enlighten* me." This was followed by another silence, more strained than before. It was anything but a repose.

"Sorry, boss, we didn't mean nothing by it," Jimmy finally said.

Jimmy thrust his head out the window.

"Boss, it's just that we're feeling pretty good."

"I can see that," Jacques said. "Well, boys, I don't mean to cut into your fun, but...ah, if I might be so bold as to ask, *why* are you gents not working, this morning? Or have you forgotten that today *is* a work day?"

There was a long pause.

Tom studied the strange expression on Jacques' face. Another mask. Was life a charade? The dark toothy hole that dominated the middle of Jacques' face seemed like some great yaw in the earth. A wild thought. Was it about to swallow the boss? Tom shuddered.

"It's a long story, chief..."

"Fire away, Jimmy. I'm not going anywhere. In fact, I've got all day." Jacques turned the key and pulled the plug. His truck motor coughed and died.

A moment later Thurston did the same. Silence embraced the two trucks, parked side-by-side in the middle of the road, facing in opposite directions. But it was not a pleasant silence. It was pregnant, urgent, insistent. The boss was waiting impatiently and he wanted his answer. Thurston knew he was at the plate. He squirmed and cleared his throat.

"Well, you see, boss, hhhmmmmggrph, it all started yesterday when...uh...we...."

Jacques sensed another raft of bullshit headed in his direction. He could feel the merciless ire rising, the old hot blood. Bile was moving under pressure of millions of tons p.s.i., bilious gall, by-passing all non-essential organs, rising up, up, up, up, this time, all the way into his throat. Jacques nearly choked, but somehow he managed to contain the emotion.

"Mu-mushrooms, boss," Thurston stammered.

"MUSHROOMS!" Jacques exploded. "WHAT IN THE PISSANT SOUR PRICK AND DIP SHIT FUZZY DAMN ARE YOU TALKING ABOUT?"

"In the woods, boss," Jimmy said. No more hesitation. *"Mushrooms."*

The word fell into the synaptic void behind Jacques' forehead, sank into the swirling vortex, and vanished. Came an eddy, a cleft pause, a moment of imperturbable silence, a dead space of non-reaction. Disbelief, maybe? Denial? No, the pause was only a punctuation, the space between thoughts, the sound of the *Nada,* empty perhaps, but infinitely fertile. Make no mistake, the word had found its mark. *Mushrooms.* Somewhere inside Jacques St. Clair's cortex sawdust was flying, yes, sawdust and wood chips and contractual fine print and dreams and plans, a life, all scattering and unraveling, insubstantial as the wind, unreal as a fanciful daydream, all slipping away like sand through his fingers, everything drifting, decaying into nothingness.

Stung to the quick, the boss knew he was beat. Hating defeat, all he could think to do was reach for the key. Fourth down. Can't make it. Time to punt. Jacques turned the switch. The ignition coughed. The engine turned over and lived again.

Borrowed time.

"I see...uhhh...OK, boys...I'll...uh...get back to you in a little bit," Jacques said with a wan smile. All color and expression had drained from his face. Sucking in his chest, the boss gently nudged the gas. Slowly the truck began to move and pull out, leaving Jimmy and the others behind, waiting in the middle of the road.

Jimmy was incredulous, his mouth open. The others, Dipstick, Charlie, Red, and Tom likewise. No snickering, now, only stares of surprise and disbelief as the chief pulled away. Looks of incomprehension.

Jacques' shoulders were hunched forward, as if burdened, his eyes fixed on the road. He allowed not a glance to the side, not even a peep in the rear view. Gently he eased his foot into it, feeding gas to the engine.

SLOWLY. Easy does it. Keep the foot light on the pedal, not too much gas, no more than a touch. That's it. Steady as she goes, light as a feather.

Finally, he took a deep breath. The truck was putting distance between, safe distance, uniformly accelerating so as to spin no gravel, leave no tracks, stir up no cloud of telltale dust. Nothing to draw attention to Jacques St. Clair's tail-door flapping open in the breeze.

Cardinal boss rule *numero uno:* never let the men see you sweat.

BOOK THREE

THIRTY EIGHT

A major storm system blew in from the Pacific bringing unsettled weather and unseasonal cold. The front piled up ominously against the Rockies, enveloping the high country in a sheath of clouds.

A sharp blustery wind was blowing as Tom drove into Granby. He meant to stop by the house where the activist meeting had occurred for an update about Bowen Gulch, but as he cruised by the Forest Service district office on Main Street he noticed a demonstration in progress. A large crowd had occupied the parking lot, and was milling around.

Damn!

A line of protesters also stood along the highway. They were waving signs about Bowen Gulch.

He parked his rig and crossed the street. A woman handed him a flier as he approached the demo. As he edged his way through the crowd which he guessed at maybe two hundred people, he noticed that a second smaller group had gathered around a vehicle in the lot. The center of curious attention was the flatbed with the huge spruce log that had been trucked around the state.

He recognized some of the faces. A small boy struggled with a sign that was a lot bigger than he was. The sign read: SAVE BOWEN GULCH FOR MY CHILDREN'S CHILDREN.

The math professor was up in front with a bullhorn, apparently done speaking. Pinecone was nowhere to be seen.

Suddenly, two men in Forest Service green came out the front door of the district office and hurried down the walkway toward the highway, actually toward the road sign, under which someone had affixed a banner. The official sign read:

Arapaho National Forest
Sulfur Ranger District

Beneath was a white sheet with bold red letters:

A SUBSIDIARY OF WESTERN-PACIFIC

The visibly unhappy staffers proceeded to rip the banner down. As they did, Dr. Newsome lifted the bullhorn. "Our Forest Service friends, over here," he said, pointing with his other arm, "Obviously do not appreciate our little wake up call. They think we defaced their sign; but, folks, the truth is, we corrected it." Cheers from the crowd. Now, he addressed the staffers directly. "You men, have you ever heard of John B. Sewell?" Newsome paused, waiting. When there was no response, he continued. "It seems our friends are not aware that in 1981 Mr. Sewell, lead attorney for Western-Pacific, was named Assistant Secretary of Agriculture, which placed him administratively *over* the Forest Service." He paused. "So, you see, we have not defaced anything, our sign is factually correct. The timber industry is not only in bed with the government, hell, they *are* the government!"

The assembly exploded with cheers and a few catcalls.

Newsome continued. "There are times, precious moments in life, when friends have to level with one another. We call it 'tough love' because the truth can be painful. Hey, isn't that what true friends are for?" More applause.

Tom finally looked at the flier. As he read through it he muttered, "Holy crap!" Night before last, somebody had torched a dozer and road grader. The St. Clair brothers were now out of the picture. W-P had ripped up Jacques' contract and brought in a new boss with a different crew. It was a local operator, Dorfman Logging. Tom already knew about Jacques, but had not heard about the latest ecotage. Someone had totally destroyed Dorfman's bulldozer with a thermitic charge, melting the engine bloc. Dorfman's road grader had also been destroyed. Someone had torched all of its tires and everything else on the machine that would burn. The tires had been reduced to smoking drek. There had also been another round of civil disobedience. Early that morning, a half-dozen tree-sitters had been arrested at Bowen Gulch. Things were heating up.

The speaker changed course. "For the record, I'd like to apologize to Mr. Dorfman, owner and operator of Dorfman Logging. Is he with us here, today? Anyone know? I don't see him. OK. Would somebody with the press, one of you, please take my apology to Mr. Dorfman. From today's paper, it appears that yesterday Mr. Dorfman's bulldozer and road grader met an unfortunate end." He paused. "I was sorry to learn about this because Mr. Dorfman is not our enemy. I happen to know the man. He is a nice guy. He's not making a big policy decision, here. He's just trying to make a living." Another pause. "I want him to know how I feel about it because I believe it's important for us to be straight with one another, and to go on record. Now that I've done that, I also need to tell you that if Mr. Dorfman's dozer and road grader had *not* been trashed, they would have pushed that logging road into Bowen Gulch, yesterday afternoon, and by now the fight to save the big spruce trees up there would be about over. So, I also want to thank the brave someone who took history into his own hands." He was interrupted by loud cheers and clapping. "But for him, or whoever, there would be nothing left to rally for, here today. That is a fact. You may not like it. But it's the truth..."

The crowd was agitated. The professor waited for calm. "Folks, we've been hearing the 'terrorism' word a lot in the press, lately. They like to throw it around, and they mainly direct it at us. I call it the 'T' word. How come, I would like to know, we never hear the word mentioned when our military napalms a village, or when our Central Intelligence

Agency overthrows some foreign government? Can somebody please explain that for me? Heck, we carpet-bombed thousands of square miles of precious tropical rainforest in Southeast Asia, not to mention using a toxic chemical by the ton, Agent Orange, and I never once heard the 'T' word, then. Not even once. Nary a peep. Can one of you folks with the press please explain this for me? Help me out, here." He paused, waiting. "What? Nothing? Nor do we hear about it from you reporters when Western-Pacific violates the Clean Air Act, as they've been doing every day for more than a year!" Newsome was interrupted again by loud cheers. He raised his voice in an effort to talk above the crowd. "Folks, if you ask me, *this* is the real terrorism!"

Apparently on cue, several young people in the crowd unfurled a large banner, obviously meant for the press. Several journalists from one of the Denver TV networks were in the crowd, with cameras. The banner read:

LOGGING BOWEN GULCH = PLANETARY TERRORISM

Suddenly, she was standing beside him. Tallie placed her hand in his and smiled at him in that way that made him feel so alive. "My aunt dropped me off," she said breathlessly. He nodded, squeezing her hand. Her face was red. "You got some sun," he said.

The demonstration continued for about another twenty minutes. There were more speakers and some announcements, after which the demo played out. When the crowd began to disperse he walked with her to a small cafe and bought her lunch.

She had been at her cousin's place in Denver when he called from the Kawuneechee Lodge, and so, had missed the round-the-clock mushroom fiasco at the logging camp. She was eager to hear the story, how Jacques' logging crew had been transformed into a bunch of besotted fungus eaters. He spun it out from the beginning, including the part about Wolfe and Bobby. She listened without comment, plainly enchanted. As he got deeper into the telling he warmed to it. The story was quite a yarn, after all, spiced with plenty of humor in the madcap details and the incongruities. He found himself chuckling all over again. He could not help himself as he described the crazy scene in camp and how Jacques St. Clair went ballistic.

Her eyes grew large as he told of the manna's strange disappearance on the third stuporous morning when an ever-widening sweep through the timber failed to turn up even a single fruiting body. The "food of the gods" had vanished as mysteriously as it had appeared. Where, before, one needed tramp only a few yards through the forest to collect mushrooms galore; earlier they had picked bagfuls, in the end, only a few half-eaten stubs remained. Nature in her caprice had withdrawn the treasure. Judging from the tracks, wild animals had come in the night and grazed them down to the nubbins; an entire year's crop gone in a single night. Other species, it seemed, relished the magical mushrooms as much as humans.

He described the crew's shock and bewilderment as collective euphoria turned to deep disappointment. Nor did Jacques St. Clair succeed in regaining control of his crew. He described the confused and angry showdown on the landing when Jacques' brother Paul arrived from Durango with the second crew: how the two brothers had argued, first with one another, then with the Western-Pacific attorney; and the subsequent near fist-fight on the logging deck. Neither Paul nor the company lawyer could fully grasp what had happened. By this time, most of the crew had already dispersed to the winds chasing a psychedelic rainbow, in search of more mushrooms. She was fascinated as he described the mushroom's aphrodisiacal properties, how the love bug had infected everyone, and how it made them horny as billy goats. But her face darkened at the rest of the story.

"Bobby had Wolfe figured right," Tom said, and related how the cop heard a muffled scream as he was about to leave the camp, then returned to one of the campers where he found Wolfe in the act of violently sodomizing the Preacher.

"Bobby got his Arapaho justice. They figure he's looking at five to ten. " Her buoyant face had deflated. "Come on, I'll take you home."

THIRTY NINE

The next twenty-four hours were a blur. It all ran together like a dream. Late in the afternoon, after he dropped off Tallie, Tom doubled back to the derelict house for a training session in non-violence. The group was an even dozen and a mix of ages, and also equally split gender-wise, half male and half female. A women named Roberta Hoss led the training. She had them sit around her in a circle. Then, as she went around, she asked each to give his or her name and explain why they had come. After the introductions, the actual training started.

It was very straightforward. Roberta emphasized that although the Ancient Forest Rescue coalition was committed to non-violence there was nothing passive about what they were trying to do. "Civil disobedience is a very active and aggressive form of political resistance," she said. At that point, she introduced a term, *Satyagraha,* the importance of which, she stressed, would be impossible to exaggerate. She explained its meaning. The term had been coined by Mahatma Gandhi, the great soul who had famously led the Indian nonviolent resistance against the British imperialists. The word was a composite of two Sanskrit words: *"Satya,"* meaning "truth," and *"Agraha,"* meaning: "to hold onto." The composite meaning, therefore, was "holding onto the truth."

Roberta presented her interpretation, emphasizing that it was essential that protesters maintained self-control at all times, no matter what, even in the face of extreme provocations by the authorities. She believed they were likely to face police violence at Bowen Gulch and that the chances of it happening would increase as the campaign to stop the logging intensified. She said she expected things might get very ugly before it was over. "And this is why," she said, "it is essential for each one of you to remain inwardly calm at all times. You have to find your own personal center of gravity, and hold onto it. You need to ground yourselves and live in that place. We view this as mandatory," she said. "We can't have anyone losing control, flying off the handle. If you feel that you are not able to do this, you had better leave now. Self-control is a prerequisite."

There were additional instructions. She cautioned that once the action started they should avoid making any abrupt moves or gestures, anything the cops might interpret as a violent escalation. For example, they should always walk, and never run. She also cautioned them never to shout at the cops, or make insulting remarks. For obvious reasons.

"It's also best," she said, "to try to maintain eye contact with the police, if possible. If you choose to engage in civil disobedience and are arrested, you can either go limp, or you can cooperate. It's your choice. But you need to be aware that if you go limp and are carried away to jail in a paddy wagon, you might be charged with 'resisting arrest.' The courts do not rule consistently on this. It will depend on the judge."

Needless to say, alcohol and other drugs were banned. No exceptions.

They finished the training with a song, which they sang while holding hands. It was a rendition of Amazing Grace, an old standard that Roberta said came out of the political struggle in England to ban the slave trade at the turn of the 19th century. She said that the struggle to preserve ancient forests was a continuation of the "good fight" for a better world waged by the best people of every generation, right up to the present day. "We should feel proud," she said, "that it's our turn to carry the torch in this great time-honored tradition." She led it off:

"Amazing grace, how sweet the sound, that saved a wretch like me...
I once was lost, but now I'm found, was blind but now I see..."

Tom thought everything he heard was logical and perfectly obvious. It was easy to appreciate the emphasis on discipline. Even the singing served a purpose. There was no way to remain unmoved, emotionally inert. By the time they finished the training they were joking and laughing with one another. A bond of trust was already forming among them. They were becoming an extended family. "It's all good," someone said.

One of the activists explained another reason for the training. It served, he said, the practical function of helping to unmask agent provocateurs. Paranoia about government spies ran deep among the activists and for good reason. Provocateurs were known to be afoot. One likely case had been unmasked a few days before. The individual, a self-proclaimed Vietnam vet, had arrived talking trash. The man always took the most extreme position on every issue, the initial red flag that made people suspicious. When he started talking about explosives and blowing things up they showed him the door. The guy was invited to leave and told not to return.

As the session wound down they could hear people gathering in the large adjoining room. From the clatter in the kitchen it was evident that some kind of meal was being prepared. Just about the time they finished, someone announced that dinner was ready. They opened the doors wide and two-dozen hungry activists filed through the kitchen as volunteers handed out bowels of vegetarian chili and cornbread served with butter, and mugs of iced tea.

Two women went out the front door with a tray to feed the Granby cop parked on the street. The policeman had apparently been ordered to watch the house; though, thankfully, he was always gone by nightfall. The activists had by now grown accustomed to his presence. They waved and greeted him with friendly hellos. Some even chatted with him. He was a nice enough guy, just doing his job.

They packed in to the large living room. Someone turned on a television so they could watch the evening news while they ate. The group was eager to hear the latest about Bowen Gulch. They were not disappointed. A Denver news channel carried a breaking story that glued them to the screen. Earlier that day, a major home developer in Boulder Country had announced that he would no longer purchase lumber milled by Western-Pacific if the company went ahead with the Bowen Gulch timber sale. In a taped interview the developer and his wife, a Mr. and Mrs. Ralph McGee, of McGee Homes, explained why they were supporting the boycott. "We just decided," Mrs. McGee said, "that as responsible builders we can not justify making wafer-board and small dimensional framing lumber out of 600-year old trees. It just makes no sense to us. So, when we got the call from Jan, with the Ancient Forest Rescue, we told her we were proud to join the boycott." A collective cheer rocked the house.

"This is sooo huge!"

"Can you believe it?"

"That'll increase the pressure on W-P."

"I never thought we'd get *this* far..."

"Things are moving our way!"

Then came the bad news. It arrived in the person of Pinecone Peters who now stood in the hallway facing them. He had come straight from Bowen Gulch. "They've started logging," he half-shouted. "Dorfman sent his loggers in about mid-afternoon. They worked until dark, dropping trees. Looks like they're going after the biggest trees first."

There was a stunned silence, then a collective moan that was probably heard two blocks away.

"What about Susan and our tree sitters?" Roberta wanted to know.

"They cleared out the last of them about two this afternoon. All arrested. That's when Dorfman sent his cutters in. They started with the lowest unit." Collective mayhem drowned out Pinecone's last words. The outrage in the room was incendiary. Everyone understood that time was up. Some of the faces showed despair.

"We're screwed."

"Shit. Those bastards..."

"Wait! It isn't over, not yet!" Pinecone shouted at them. "We're a long way from done." That is when he noticed Tom sitting quietly in the corner. "Well, well, well." The hubbub continued. "Hey! Listen up. Quiet down! I want all of you to.... Hey! Come on! Cool it!" He waited a moment. The noise dropped a decibel. "Friends, this is a special moment. Guess who's with us? He's been sitting here among you, and you didn't even notice him. The man of the hour."

"What are you talking about?'

"Allow me to introduce..."

"Who?"

"Man of the hour?"

"By now, all of you know what happened to Right of Way Inc.? Right? Well, I want you to meet the man who pulled it off. The guy who shut down the whole friggin' operation – took out an entire logging crew with one fell swoop."

The room was finally quiet. Pinecone had their full attention. He motioned with his arm. The room of faces turned toward Tom.

"Stopped 'em dead in their tracks, a hero I tell you, and none of you recognized him..." Pinecone laughed.

Tom was uncomfortable in the spotlight. He knew he should say something. "Heck, it wasn't my idea."

"That's not the way I heard it."

Tom was confused. How did Pinecone know about the mushrooms? Surely not from the loggers. "You don't know Bobby?"

"Course, I know him. Grew up with the bastard. Crazy Indian." Pinecone was laughing as he moved across the room. He had to wade through a sea of bodies. Tom got to his feet. "Man, that was a nice piece of work. I figure you bought us two days."

A woman looked up. "Two days? How?"

Pinecone laughed. "Magic. Right Tom? With a capital 'M.'"

"Magic?"

"What do you mean?"

"Mushrooms," Pinecone said, laughing. Some of the faces lit up. But others were still blank. They did not get it.

"It wasn't much."

"Yeah. Right. Look at him. Mr. Aw Shucks. Two precious days he bought us." Turning, he addressed the room. "Tonight is the night, folks. We are going to need all of you. Every last warm body we can muster. We are at the tipping point."

FORTY

After Pinecone left they cleared the big room for another training session. Fresh recruits were still arriving. The house was stuffy and someone began opening windows. The green chili fart-fest was underway. "Open them up quick!" someone cried, "Before the place explodes!"

Tom was happy to step outside. It had turned cold. There was a sharp bite to the night air that felt good against his face. It was beginning to drizzle. He went to his pickup, parked out on the street. He was thankful for the camper shell; but the back was stowed with gear. So, he had to move his saw and other stuff into the cab to make room so he could spread out his bedroll. He lit the lantern. He was wide awake and thought he might read awhile. He had several books from Mary's library, including Thoreau's essay, *On Civil Disobedience.* However, he kept staring at the page; he had too much on his mind; Tallie, for one thing, and the things she had told him over lunch. Finally he gave it up, doused the light and lay in the sack listening to the rain on the roof.

It was coming down much harder now, a steady drum beat to his thoughts.

He had come a long way in a short time. He was in flux. Everything seemed to be happening all at once. Ten days before, he had been dropping big trees without a thought. Ho hum. Just another day on the job.

He recalled the hike into Bowen Gulch on the morning after the mushroom free-for-all.

Failing to reach Tallie by phone, he decided the time had come to see this place that everyone was talking about with his own eyes. By then he was losing interest in mushroom mania in any event, especially the group-think that never held sway with him. He was too much of a loner to follow some bemused pied piper up or down a yellow brick road in search of nirvana. That was so much nonsense in his view, pie-in-the-sky. Let the others do whatever they wanted, great, fine, but it was not his style. He went his own way.

Of the solo trek he told no one, not even Tallie. The tramp into the deep forest cut too near to the bone to share, even with her, for it was more than a hike; it was a journey on seldom used pathways into his own shadowy hinterland, to poorly known places and parts within himself where, if luck was with him, he might render up some meaning, if not personal redemption, out of the reliquary of his own private purgatory.

From the landing he worked his way up through the lower sale units. The going was rougher than expected. For awhile he followed the yellow ribbons. The girth of the trees was astounding. Some of the shaggy-barked spruces were at least four feet across. He had never seen trees of this size. He knew they had to be incredibly old. The ancient trunks were covered with curlicue-shaped yellow-green lichens and moss of various kinds. He crossed a little creek, left the flagging behind, and wandered at whim.

On every side enormous spruces and smooth-skinned firs receded into shadows. He saw no yellow bellies but this was not surprising. He knew he was at least ten thousand feet in elevation, too high for Ponderosa pine.

As he moved deeper into the old growth his thoughts fell away and his senses awakened to everything around him. The air was rich with the pungent smells of decay. Here nothing was wasted. Everything was recycled. He began to feel the wildness of the place.

He continued up valley for about a mile, heading generally west, and re-crossed the stream several times.

Here and there he paused to examine individual trees. Each had its own story, if a man had an eye to read it. One huge bole had been shattered up top, its uppermost crown blown out long ago, probably during a lightning storm. The event had been recorded in a long vertical scar spiraling down the trunk all the way to the ground. Despite the damage, a leader had sprouted from just below the high gash and now pushed skyward – an indomitable act of bright green defiance.

Another ancient tree was festooned with mistletoe, its upper crown a crazy quilt. Seven feet off the ground a bear had once sharpened its claws on the trunk, an enormous paw judging from the width between the deeply etched grooves. This was no black bear. It had to be the centuries-old mark of a passing grizz, long extirpated from these parts. The tree was marked for cutting, probably because of its extreme age and decadence. A chainsaw was about to erase a last piece of natural history, just as the long rifle had the bear.

A Stellars Jay squawked loudly at him from a low branch as he studied a column of ants ascending up another nearby ruff-barked trunk. Their count was inestimable but had to number in the tens of thousands. Craning his neck, he followed the slowly moving caravan up to the limit of his vision. What their business might be in the high crown he had no idea. Another mystery.

He was amazed by how lush everything up here was. As he climbed higher there were scattered patches of snow.

Presently though he found himself traversing a slope with a drier southern aspect and he noticed fire scars on some of the trees. He paused to admire a large spruce that was nothing if not a masterwork of biological art. Some ancient conflagration had worked its will upon the tree. The trunk had been gutted, completely hollowed out. The fire-blackened cave, open to the south, now held court on a profusion of forbs and wildflowers. Stepping inside the garden cave he looked up through the bore and saw blue above, open sky. Except for the chamber at its core the spruce appeared to be in perfect health, sound enough to last for centuries. But he saw few other fire-scarred trees. Most of Bowen Gulch was too wet to burn.

He worked his way up the edge of the high valley until he came to a spring, in a kind of natural grotto. Thirsty from the hike, he knelt and drank his fill. The clear water was very cold and pure. Suddenly hungry, he pulled out the cheese and tomato sandwich he had packed along. It was delicious.

The grotto was a peaceful spot and he lingered. He was sitting on a rock in a contemplative mood when he heard a noise and looked up, in time to see an orange-brown blur come sliding down the mossy slope behind the spring. The next moment he was face to face with the most formidable creature he had ever encountered in a coat of fur; a full-grown badger not more than ten feet away.

Too close!

The badger had come down for a drink belly-first, riding his fur like a toboggan. Without a sound the critter lowered its nose, bared its teeth, and stared him down with beady eyes of steel that left no doubt who owned the spring. Tom rose slowly and backed away without ever taking his eyes off those fearsome teeth. He did not doubt that discretion in this case was the better part of valor. Chalk one up for Mister Badger.

Buddy, it's all yours...

There was more snow now as he pushed up the draw and came to a wet meadow. Here, water was everywhere on the move. The ground was a broad sheet of fast-moving snowmelt. He searched for solid footing, some way around, but there was none, so he slogged ahead through muck and mire. When he reached the far side he stopped. An empty Coors beer can lay on the ground at his feet.

Right. Just what the world needs.

He picked it up, crushed it and stuffed it in a pocket, then, moved on.

It was the deepening snow that finally stopped him. When he reached a snow-bank that was up to his thighs, he turned and started back. At some point, he reached the flagging for the skid road, and followed it down valley. He lost track of time.

*****

As a member of the State Board of Agriculture, Tallie's uncle, Bernard Holloway, had known the late Dr. Nolan Leadbetter. "I knew him well," Bernard told him. "Nolan and I were good friends. We went way back to the days when he was a graduate student, before he made a name for himself. I was working on my doctorate in those days."

"In zoology?"

"No, my boy. Electrical engineering. Both of us were members of Alpha Kappa Lambda. But the frat house was just a passing phase. We soon outgrew the Mickey Mouse club, the rah-rah business, the food fights and the hazing. All of that adolescent nonsense. We moved on. Nolan had a love for learning that I found inspirational. Truly. The man was brilliant. I have never known anyone so excited about his chosen field..."

"He was the best professor I had at state. By far. The only one in four years I truly respected."

"I'm not surprised to hear it. Nolan had a flare for teaching. He was a natural. I think some people are born with the gift. He had the bug and he was contagious. Even in the early days it was obvious that Nolan would go far. It's why people were so shaken by his tragic suicide. It was just...so unexpected. I was shocked too, of course, like everyone. But I must say I wasn't surprised. No, not really."

"No? Why not?"

The man paused for a moment, as if to choose his words. "You see, Nolan felt that he'd been upstaged, and well, he just couldn't handle it."

"Upstaged? By whom?"

"By his wife, of course!" Bernard said. "Who else?"

Tom was speechless. "But...how?"

"That's my question for you, Tom. How well did you know them? I mean the two of them."

"Very well. At least, I thought I did. I worked with Leadbetter and spent a lot of time in the zoology lab. He was my adviser for two years."

"How well did you know his wife?"

"Henrietta. I knew she taught psychology."

"Did you know about their competition?"

"What? No."

Bernard stared into space, as if remembering. He sighed, then, continued. "Nolan and Henrietta had one of the strangest relationships. I knew them both quite well, you see. I was Nolan's best man at his wedding in '58. We were drinking buddies in those days. Nolan and Henrietta had a kind of weird competition going between them. Oh, I imagine it started innocently enough. Good naturedly. But I think gradually it changed into something else, something unhealthy and horrible that killed my friend. There was a kind of strange antagonism in their relationship. An undercurrent. Both of them were professionals and I can tell you they did not marry for love. I know this for a fact. I suppose you are aware that Leadbetter published a book about the mammals of Colorado. But did you know about Henrietta's?"

"Didn't she publish a book on psychology?"

"Yes."

"I didn't know it was such a big deal."

"Oh it was a big deal, alright, a very big deal, from Nolan's standpoint. Did you know that his wife's book became a standard college textbook?"

"No."

"It did. In fact, the book is still in wide use. By contrast, Nolan's volume about the mammals was poorly bound and edited. It was full of typos. He was very dissatisfied with it. The book soon went out of print. Nolan was his own worst critic. I just didn't know the degree of his unhappiness. Instead of publishing the definitive work in his field, which had been his intent, he flopped, or felt he had; and knowing Henrietta as I do, I'm sure she never let him forget it, not for one moment. Looking back, I'd guess that Nolan was clinically depressed toward the end. A good marriage, Tom, can bring a man the greatest happiness on this earth, but a bad one can destroy him. The good Lord just never intended for husband and wife to compete."

"But he didn't seem depressed."

"Oh, he masked it quite well. But understand, Nolan was all about achievement, and on the level that mattered most, he felt that his career, hence, his life, had been a dismal failure."

"But he was a great lecturer. How could he think that?"

"Oh, I know. I agree. Nolan was a genius. *But what good was that to him?* Did that save him? No. What the man needed was a counselor, someone to confide in, and talk to. Unfortunately, that option was unavailable to him. He could not confide in his wife because of the competitive nature of their relationship; and because her domain was psychology he felt constrained. A therapist was out of the question. They were both so territorial, very turf conscious, if you know what I mean. Nolan was also proud. There was more going on than you knew, Tom, more, I'm afraid, than you could possibly have known..."

*****

He gaped at the enormous carcass at his feet. Someone had just dropped a huge spruce tree, at least thirty inches in diameter, even bigger on the butt. The cut face was raw. A pile of fresh sawdust and shavings lay beside the stump. The air was fragrant with conifer.

He stared at the fallen giant like a man under the influence.

A thousand times he had dropped trees no different from this one. How many times, during the last year, had he heard someone say, "Working in the woods is the best work there is, Mick, the best an honest Joe will ever find." He had heard all manner of men say that and a dozen similar things, yes, more times than he could recall. He had heard it from the likes of Wolfe Withers, and from mental midgets like Shorty Dibbs; but he had also heard it from regular fellows, men with attractive wives and adorable children. He had heard it from snot-nosed loggers who liked to pack a cheek-full of Skoal up against the gum, and from guys who were in the habit of fingering the waxed end of a handlebar mustache while they bullshitted about nothing in particular during lunch break. He had heard it from men who agreed on nothing else and he had listened to them all.

Well, why not?

Why not indeed? Was it not God's own truth? Were not such men the salt of the earth? Surely they were at least as honorable as the lawyers and car salesmen and bankers who screwed people for a living, and the professors who put you to sleep in the name of higher learning. Men who could not even change their own car oil, who would be lost trying to fix a faucet, who would likely pound their own thumb into hamburger if they attempted the simple task of driving a nail.

Those who can, do; those who can't, teach...

What could be more natural or right than dropping a big one for love or money? Was it wrong for loggers to take pride in their work? Was it wrong to enjoy watching the big trees fall? And if logging was the best paying work an uneducated man like Shorty Dibbs would ever find, so? What of that? Who had the right to say it was wrong? Sure, it was about money. Of course it was, but the same could be said of any other trade. Was not logging just another way for a man to feed his family? A man had to take care of his family. Besides, did the country not need timber? Wood for boards and beams to make homes so people could realize their dreams. People needed timber for joists and rafters and processed doors, and veneer and plywood. The country needed paper too, cardboard, furniture and a thousand other things. Did not wood products keep the economy humming? Was it not timber as much as oil that made the wheels go 'round? All true. So what was special about this particular spruce tree? Why did *this* downed tree feel like a violation?

Pathetic...

Tom looked around. The tree was not marked, nor was it even within a sale unit. He saw no yellow flagging.

So why drop it? Why would someone hike into a roadless area with a chainsaw?

He did not like the answer.

Probably for no reason at all. Probably for the hell of it, just to watch it fall.

What finally brings a man to the point of change? What tips the balance in the end? Is it fear, or love? Certainly it is not money. Do the songbirds sing a bit more sweetly on such a day? Is the sky more blue, the sunlight brighter? Is it something in the water or the air? Or is it something in the man?

Tom found himself listening to a woodpecker as he drifted with his thoughts. He loved that drumming sound. Looking around, he could not locate the bird. The woody was back there in the timber somewhere.

Just as all things have a point of origin, so too when a mind at the end of its tether unaccountably snaps its cord and sails into parts unknown, a new idea alighted in the cavity between Tom's ears, like some downy woodpecker. It was a novel thing, this contemplative bird, and all the more surprising because apart from the space in his head it had no other home.

He found that he was staring down at the stump. Clear sap was still oozing out over the raw face, seeping up from deep below ground.

Dead. Yes. Dead. But the tree doesn't know it, yet...

Things were all mixed up, a mess of disordered impressions.

Suddenly, he wanted to strangle the stupid son of a bitch who for no reason at all had dropped this big spruce. He wanted to shove the guy's face down down down, into the sap and show him the stupid thing he had done, to make him see, make him understand what a damned fool he was. The next moment was the worst.

It might have been me.

Was he so different from that guy? Was he any better? No, in the final analysis, he was pretty much the same as the dumb bastard who had dropped that tree.

For an agonizing moment Tom was bereft. Self-loathing turned to shame such as he had never known.

Where does a man turn for solace in such a dark moment? Where does he turn for strength? If he has an ounce of sense, he goes within.

Tom's hands had tightened into hard fists. There was no point in dithering. He turned to go.

He had not gone 200 feet before he nearly tripped over a surveyor's stake. Without a thought he pulled it up and heaved it back into the forest. He did the same for two dozen more as he worked his way down to the landing. He also collected every yellow ribbon that he saw, and stuffed them in his pockets.

That was when he noticed the springy feel of the duff under foot and his own light step. He began to hum and sing an old Johnny Paycheck number. He remembered only a few lines and kept repeating them over and over:

Take this job and shove it! I ain't workin' here no more.

FORTY ONE

He was awakened by sharp knocking on the canopy window. He thought of Tallie, but this was different – very insistent.

"You Tom Lacey?"

"Pinecone?"

"No," said a voice. "Steve Gaylord."

The tail door flew open and the next moment the man was shining a bright light in his face.

"Pinecone sent me. Chop-chop. Shake, rattle and roll. We leave in five. Bring a flashlight and dress warm. It's supposed to rain all night." The guy moved on to the next vehicle, a camper where he rousted another warm body. Tom found his long johns and slipped on an extra pair of socks. He dressed in layers, grabbed his poncho, and went out the back. It was coming down hard now, in sheets, a cold driving rain. The night was pitch black but he could see his breath. He covered his head with the hood of his poncho. A van was parked in the street with the parking lights on. The motor was idling quietly. He opened the rear door. It was warm and steamy inside. Two faces stared back at him.

"Hey," said one.

"Morning," said Tom and climbed in. "What time is it?"

"Don't ask," a woman said.

"What the fuck difference does it make?" the other face said, a man with a ragged beard.

He found his spot. The rear door opened again. Four more climbed in. Two he recognized from the training. Then, the front door opened and slammed again. The driver was back. He looked over his shoulder.

"Everyone alive?"

"Yeah, let's go."

"Alrightee. Next stop Bowen Gulch." Steve Gaylord popped the clutch. The vehicle lurched forward.

It was dark in the van. The only light was the pale green glow of the dashboard. Tom's eyes adjusted. They were packed in, six bodies on the floor in back, two rows of three facing one other – plus the two in front, a total of eight. They bounced and swayed together as the van made a sharp turn out onto the highway. The driver picked up speed. They rode in silence. Apart from the engine the only noise was the flapping of the wipers on the windshield. It was easy to imagine they were soldiers on the way to the front. One of the other activists also picked up the vibe, a free-floater, because a voice said, "It's a good day to die."

"Right on," said another. "Today, we take no prisoners."

Then the conversation took off like a fireball out of hell. It was straight ahead no-bullshit politics. These folks had an attitude and were not a bit bashful telling you about it. They were unanimous in their disapproval of "those lame liberal politicians" who were "even worse than used car salesmen and corporate attorneys." And why? Simple. Because "In the clutch a lib'ral will betray you, every time, stab you in the back and leave you to twist slowly." Lib'rals were "incapable of right action, each one with his own sorry excuse." And: "Count on a weak-kneed lib'ral to blow over in the slightest breeze."

Tom was by now feeling something bordering on sympathy for these folks he did not know from Adam. He had never thought about it all that much, but he knew he was no "lib'ral." Some of the stuff he heard put him off a little. Several of the activists were brash to the point of being reckless. Even so, he knew that somehow he liked and trusted them. At any rate, he welcomed their no-nonsense attitude, which he found refreshing. Their blunt honesty was like oxygen.

Having ranged through general themes the conversation now turned to particulars. They discussed tactics. The group was obviously accustomed to a type of democratic process. It did not take them long to reach a consensus. The issue at hand was quickly decided. They would make a stand at the Forest Service gate near the roadless area boundary. From a tactical standpoint the gate was more than a gate. It was a potential bottleneck. If they could hold the gate they might prevent Dorfman's crew from reaching the work site. The other objective was to block the logging trucks. They estimated that come morning, first thing, Dorfman would send out one or more loads. The drivers usually arrived for work shortly before dawn; which explained the ungodly hour. They would beat the drivers to the gate, secure it, and lock them out. Someone did a time check. It was 3:30 A.M.

They figured they could probably hold out for the better part of a day, maybe longer if enough support in the way of warm bodies materialized. They would hit them in waves. Everything depended on numbers. This group was the tip of the spear and would lead the action. However, at that very moment other activists were mobilizing in Denver, Boulder, and in other Front Range towns. Reinforcements would arrive later in the morning, though no one knew exactly when or how many. The action would end when they ran out of troops, cannon fodder; volunteers ready and willing to go the full nine yards. They joked nervously about the good time they were going to have in the county jail.

"We'll have a party!"

"Yeah, a victory celebration!"

This made Tom edgy. He was not sure he was ready for jail-time. He had no experience with civil disobedience and had never been arrested in his life. It felt like a big first step. The possibility, even the likelihood that in just a few hours he was going to be incarcerated made him think. He visualized himself stepping out of an airplane into the wild blue yonder, dropping while he prayed for his chute to open. The prospect was unnerving. Frankly, it terrified him.

Am I ready for this?

As the discussion moved forward he searched within. For what? He was not sure. For the wherewithal? Courage. Yes, and for strength. He was still searching when Steve Gaylord turned off the highway onto the secondary road to Bowen Gulch. Gaylord pulled to a stop and killed the engine. The talk died with the motor. They waited in silence in the darkness, listening to the storm lash the roof. A hard rain was, indeed, falling.

They did not have long to wait. Two other vehicles appeared. The headlights turned as one slowed and pulled alongside. A pickup. The other was a van. Gaylord rolled down the driver window and spoke with someone in the other rig, then rolled it back up and restarted the engine. The two other vehicles led the way, accelerating up the access road. Gaylord followed. It was a winding road and at some point Tom knew they had left the pavement. The feel of tires on gravel was unmistakable. It was too dark to see out the van's back window, but through the windshield he caught glimpses in the headlights of trees along the road of a wall of conifers.

Fifteen minutes later they reached the gate and emptied out. The scene was crazy with so many flashlights; twenty activists trying to get organized in the pitch dark and pouring rain.

It was decided that an Earth First! affinity group of ten would lead the action, the first wave. They were gender neutral: five men and five women. Together they swung the steel gate closed and padlocked it shut. Pinecone pulled a handful of cables out of a knapsack. Each of the ten proceeded to lock themselves to the closed gate, each one separately. When they were locked in securely they made themselves as comfortable as possible, and settled in to wait. Other activists draped them with blankets and ponchos to keep them warm and as dry as possible.

The cold rain came down in buckets.

The rest, including Tom, would provide support, whatever was needed, and also serve as witnesses when the police came and began making arrests. One of the women in the group began to sing a folk song, one Tom had never heard:

I have dreamed on this mountain
Since, first, I was my mother's daughter
And you can't just take my dream away...

She had an amazing soulful voice and sang so beautifully that when she finished the rest began howling like coyotes. Someone flashed a beam on the Forest Service sign near the gate. It read:

ROAD CLOSED TO MOTORIZED USE
LOGGING TRUCKS ONLY

Several others now also focused their beams on the sign as two men draped a sheet over it and somehow attached it from the back, or maybe the top. The sheet turned out to be a home made sign. When they finally unfurled it the new sign proclaimed in big bold letters:

THE ROAD STOPS HERE
EARTH FIRST!

There was a wild cheer and more yipping.

"That's an improvement!"

"Right on, sister."

They were ready. There was nothing now to do but wait. It was miserable in the cold rain. They covered up as best they could but there was no relent. Some of the activists began stamping their feet, to drive out the numbing cold. One of the women handed out oatmeal cookies. Roberta Hoss appeared with a large thermos and began handing out paper cups of steaming coffee. The cookies were chewy and sweet. The hot coffee was heaven sent.

Tom's thoughts were in disarray as he moved away from the group. He needed to be alone to think. He was still trying to sort out the disquiet in his soul, still searching for the intestinal fortitude that he truly doubted he possessed. Something else was bothering him as well. He needed to get his bearings, some perspective. Something about the place just did not look right. It bugged him that he did not recognize the gate, the site of the action. He had driven this road the day he walked the gulch. But nothing looked familiar in the dark and it troubled him. He moved to the fence at one side of the gate. Reaching through the strands of barbed wire, he set his cup of coffee on the ground. Then, he spread the strands apart and, bending down, ducked between them. He got through OK without snagging his poncho or trousers on the barbs. He retrieved his coffee, then, started up the road to the logging site, away from the protesters.

It was too dark to see much of anything. He swept his beam back and forth. This side of the gate, the road abruptly narrowed into a logging track, about half-mud. The gravel ended at the gate. There was no shoulder, just deep forest on both sides. The road climbed steeply. Stopping, he now had his fix. He knew exactly where he was. It was weird how things looked so different in the darkness and rain. He knew that the landing, where Jacques had parked his office trailer, was at the top of this same hill, less than half a mile away. The two disabled skidders were probably still parked there too, where Jacques had left them.

That was when he noticed the tire tracks in the mud. He knelt down and studied them with his light. He could see they were fresh, despite the rain. Several vehicles had already gone through, up this very road, *shortly before the protesters had arrived*. The group had missed the tracks on the other side of the gate, probably because of the gravel. No mistake, Dorfman was a shrewd operator. Evidently the man had anticipated some type of action this morning and intended to outmaneuver it, to beat the protesters at their own game. Tom now understood.

The logging trucks are already on site.

Dorfman probably meant to send out one or more loads before dawn. This could happen at any time. He wheeled around at the closed gate which was now about fifty-sixty yards behind him. He understood.

Jesus!

Even if the driver of the first rig out was paying attention he would probably fail to see the protesters in the rain.

Or if he is going too fast!

The gate had been wide open when the drivers came through. They probably did not even notice the gate on the way in.

They sure as hell won't be thinking about it on the way out! On this steep slope they won't be able to stop. Not in time! Not with a full load! Not in the rain! Not with the mud! If the driver brakes too late he'll skid right on through the gate and...!

A horrifying image of mangled bodies flashed before his eyes.

He heard the low rumble. There was no mistaking the sound of a diesel engine. A logging truck had just left the landing, no doubt fully loaded. He heard the engine accelerate. The truck was coming...

Oh shit!

He turned back to the gate and tried to think what to do. It was all happening too fast. There was no time to return and warn the protesters. They would never get all of the activists unlocked in time. Not a chance. He had to do something. But what? He waved his light frantically, screaming, "Open the gate! Open the fucking gate! A logging truck is coming!"

None of the protesters looked up. They could not hear him in the rain. It was apparent they had not heard the truck either. Now he understood why.

They are expecting trouble from the other direction.

The activists were unaware of the danger behind them. Their attention was focused the other way. Tom fought off panic. He turned uphill again and now saw the headlights blinking through the dark forest. The road curved away as it climbed. The logging truck was coming around the bend. It was still out of sight except for the headlights peeking through the forest. He waited, listening, hoping to hear the sound of gears. Truckers normally downshift for safety before they start down a steep hill, especially with a full load.

What he heard next sent a chill through him.

The driver is picking up speed.

He was accelerating! The idiot was coming way too fast. He was cannon-balling.

Tom dropped the coffee and began running uphill toward the truck, away from the gate, as fast as he could move. He knew time was short. He slipped in the mud and went down cursing in a heap. He got up and started again. He was already out of breath. Now, the headlights were sweeping around the curve. Another second and he would be in both beams.

But running was useless, he knew. He stopped in the road frantically waving his arms and his light. The truck was coming really fast now and still picking up speed. He knew he had to stay in the headlights to the very end.

Suddenly, the rig was upon him. The cab towered over him. He knew that the driver had not yet seen him. The truck was still accelerating. He stood flailing in the beams. They say time is relative. In the next moment all of Tom's hopes and fears coalesced into an eternal blink. The last thing he saw was the horror on the trucker's face. Tom surrendered to fate, threw himself away from under a huge wheel. The left front tire clipped his boot as he went down, before he hit the road. As he rolled the enormous rig passed in a blur of mud and stones. After another eternity he heard a loud screeeeching sound. To Tom it was like a sonata from heaven. The driver had awakened at last to the closed gate and the calamity looming ahead of him. The huge rig lurched, lurched again, then, still again, as the driver worked the brakes to avoid a lock-up. Almost imperceptibly, the logging truck began to decelerate. But the odds seemed too great. The distance too short and the speed ... oh the speed.

Suddenly more screeching and the engine roared – RAT-TAT-TAT-TAT-TAT-TAT-TAT

Then, the sound of an idling diesel.

When Tom reached the gate the driver's door stood open; the driver was already out and talking animatedly to the protesters. The logging truck had come to a dead stop less than a foot from the gate. The air had an acrid smell and a sound of hissing steam.

"Thank God for the Jake brake," the driver said, smiling. "First time I ever had to use it." He seemed weirdly energized by the close call.

The protesters looked stupefied. Unable to escape due to their cables, they had been helpless, like deer in the headlights.

Tom pumped the man's arm, full of admiration. He was surprised by how the guy took it all in stride. "What? Oh hey, no big deal. Was that you in the road waving your arms? Buddy, that extra few seconds made the difference."

Was it all a dream?

The rain had slowed to a drizzle by the time the deputy showed up, shortly after 8:00 A.M. The scene was chaotic. By then the access road was backed up with vehicles. A dozen irate loggers were waiting for the gate to be cleared, so they could get in and go to work. There were heated words, but not much really happened. Nobody wanted to fight in the rain. It was a Mexican standoff. The loggers mostly stood around smoking and gawking at the women protesters, some of whom were "real purty."

About 9:30 A.M. the press arrived with a camera crew and started doing live interviews.

Deputy Joe Ramirez looked frustrated and a bit overwhelmed. When the Sheriff finally showed up the deputy walked over to greet him. "Did you bring the cable cutters?" he wanted to know. He had called in the request two hours before. The poker-faced sheriff just nodded and motioned to the back seat. Ramirez retrieved the tool and returned to the gate. After that, it only took about fifteen minutes to take the ten protestors into custody. They were cuffed one at a time and walked to a waiting police van. A few minutes later, they were taken away. The deputy cut the padlock, opened the gate, and waved the big rig through. He played traffic cop until the snarl of vehicles was sorted out. The loggers climbed into their trucks and drove on through to work. Finally, the deputy noticed the protest banner, and pulled that down too.

The sheriff ordered the remaining protesters to disperse. Immediately. He warned them that if they were not gone when he returned in ten minutes, he would arrest every last one of them. Then, he and the deputy climbed into his patrol car and followed the loggers up the hill to the site, apparently to check things out at that end.

Twenty minutes later when the cops returned, they found the gate barred and locked shut, all over again. Ten more tree huggers were cabled to it, Tom among them. The second wave had struck. The sheriff and his car was on the wrong side of the gate, locked in. His face was beet red as he got out of his rig and shouted at them. "Open the damn gate!" he demanded. "Now!"

The activists refused to budge. They were hunkered down, ready for anything; whatever. Bring it on.

"I order you in the name of the law to open this gate!" the sheriff roared. "Pronto!"

No response. Not even a kumbaya. The top law enforcement officer in Grand County began pacing with his hand on his pistol, swearing under his breath. He looked pissed off beyond all measure. The deputy, meanwhile, got on the radio and called for another paddy wagon.

While they waited, another backup unit arrived on the scene, lights flashing.

Eventually, the cops brought in a mini-bus to handle the second wave. However, it did not arrive for more than an hour. By then, the sheriff was so angry he was speechless, apparently aggravated because he had missed his breakfast.

Finally, Ramirez and the other deputy started cutting the cables and making arrests. To speed things up they backed the police bus up almost to the gate. They had processed about half of the protesters when the third wave showed up, two more vans-full led by Pinecone, about twenty more tree huggers. Earlier, Pinecone had returned to Granby to ferry more in.

"You're all under arrest!" the sheriff screamed at them.

At that moment, the cop driving the bus said, "You got a call, chief."

The sheriff half-turned. "What now?" he said angrily.

"It's dispatch. They want to talk to you."

"You take it. I'm busy."

"I did. But..."

"Ask them what they want."

"I did, chief. They won't talk to me, they want you." The sheriff shrugged and stepped around the side of the bus. By that point, the two deputies had just finished arresting Roberta Hoss and were cutting Tom free from the gate. They roughly hustled him into the bus. Only two protesters remained.

The sheriff returned with a mask-like expression, a look on his face that words cannot describe. But take 'disgusted' to the nth degree and you might come close.

"Let 'em go," he said wearily.

The two deputies looked up. "What?" said Ramirez.

"You heard me."

"Chief? What do you mean?"

"I mean, Joe, let 'em go. Western-Pacific just halted the sale. They agreed to a buy-out, or some damn thing. I don't know the details."

"But we can't just turn them loose. These people broke the law. Hell, they are terrorists."

"The Forest Service is not going to press charges," the sheriff said wearily. His gruff voice was full of resignation. Plainly he was exasperated. "I can't hold them if they won't charge 'em. There's no point even booking 'em."

"But boss..."

"That was the governor on the phone, son. You want to backtalk him too? Let 'em go, I said."

A cheer went up. Activists began shouting and jumping up and down, dancing with one another and with the deputies, even with some of the loggers who were caught up in the contact high of so much excitement. Several men clapped the sheriff on the back and called him a "Good old boy!" Others lifted their arms, fists high. One of the activists pulled out a flask and started passing it around.

"What do you have, there?" one of the women demanded.

"Brandy, Jill. Have some..."

"But I thought we were alcohol-free."

"It's medicinal, for emergencies."

"Look," someone said, "it's breaking up!" A dozen heads turned to the West. It was true. The sky was clearing. A jagged patch of blue had opened over Bowen Mountain, a gap in the clouds, like a crown. It was an azure blue, the palest of blues, and it held the promise of better days.

FORTY TWO

A blustery wind was rocking the cottonwoods as Tom drove up the gravel drive. He exited his rig and paused for a moment to take in the mountains to the east. Though the high country was still wreathed in clouds the view from the front yard was breathtaking. Mary had seen him from the kitchen window and met him at the front door.

"Tom. Come in. Come in," she said, wiping her hands on her apron.

"Is Tallie home?"

"Sound asleep. She had a rough night. I'm letting her sleep."

"A migraine?"

"Thankfully, no. Not a migraine. But as you know she has good days and bad days. Yesterday was a bad one. She just needs to catch up on her rest. Are you hungry?"

"Famished."

"You came to the right place." She paused to let in Rough and Ready, who were wagging at his heels.

"Smells really good."

"I hope you like turkey."

"Love it."

"You timed it perfectly. I just took a fifteen-pound bird out of the oven." The troupe flocked into the kitchen where the canines waited patiently as Mary opened two large cans of dog food. She led them out into the laundry room and set down the bowls. The waggers got right down to business. "Let's give the turkey some time to cool. Then we'll eat. The potatoes and beans from the garden are done." She motioned to the covered pans on the stove. "Think you can hold out for half an hour?"

"No problem."

"There's chips."

"No, I'm good."

"How about a cold beer?"

"I would but I'm still damp and cold from last night. I came straight from Bowen Gulch. Last night, we about froze in the rain. Do you have something hot?"

"How about tea? I was about to make a cup for me."

Tom nodded.

"It stormed hard here too. Heavy rain and wind. I want to hear everything."

Mary put the kettle on. "Tom, how about building us a fire? Now you mention it, there's been a chill in the house all morning."

He felt the temperature drop as he left the kitchen. Kneeling before the stone hearth, he stuffed old newspapers under the wrought-iron grate, then, laid on kindling and some larger pieces. He touched it off and fed the blaze until it was roaring. Within minutes the large room was warming up. He waited by the high window enjoying the view.

Mary joined him and set a tray on the coffee table. She moved to the fire and stood sipping her tea. "Ah, that feels good," she said, "There's nothing like a wood fire." She settled herself comfortably in a rocking chair and tossed a newspaper at him. It flopped on the floor at his feet. "Did you know you made the front page of the *Denver Post?*"

"No."

"The story is short on details. Maybe you can fill me in."

At a glance he saw it was yesterday's paper. The headline read: LOGGING EQUIPMENT DESTROYED! The story was old news, but he was intrigued by the color photo of Jacques St. Clair on page one. The boss looked so different without his hat. The story was about the sabotage and the spiked trees. According to the article, the evidence pointed to environmental extremists. However, as yet no one had been charged or even apprehended. A sheriff's investigation was still underway. The paper quoted a spokesperson for the Ancient Forest Rescue coalition who decried the ecotage, denying any involvement. The article also hinted at "strange goings-on" at the logging camp, and even mentioned that some of the loggers had refused to work. He chortled. The story continued on the back page with a photo of the sheriff's deputy. Officer Joe Ramirez appeared to be scratching his head. The article raised more questions than it answered. Tom was smiling when he looked up. "Tallie told you about the mushrooms?"

"Yes." There was a curious light in Mary's eyes.

"Mary, this is old news. The war is over. It ended early this morning at Bowen Gulch. We won. We stopped the cutting." He laughed.

She stared at him intently. "We?" she said smiling. "So, when were you planning to tell me about it? Next week? If you don't get on with it I might die from curiosity."

"Where to begin."

"How about any place?"

So he did. He told her about the nonviolence training, the ride in the van, the showdown at the gate and the arrests, and how everything had played out. Mary listened with rapt attention. When he was done she said, "It would appear you just put yourself out of work."

"Something like that."

She had been studying him intently. She got up and shuffled across the room. Her slippers went *flop-flop-flop* on the flagstones. She began exploring the bookshelves, poring over several of the stacks. She was looking for something. "So, what will you do, now?" she said. He was still considering how to answer when she added, "Ah, here it is. I knew it was up here somewhere." She was on her tiptoes, reaching, and with two fingers extended just managed to slip a book off the top shelf. It was a slender green volume. She opened it with obvious relish. Several of the page corners had been folded back. A dog-eared bookmark protruded from one end. She wet her finger and began turning pages as she crossed the room. When she found what she was looking for she handed him the open book. "Is this it?" she said.

The next thing he knew he was staring at a full-page color photograph of a bright red mushroom. He flipped to the title page, and read:

THE SACRED MUSHROOM
History, Biology, and Fungal Lore
By
Cormac O'Sullivan, PhD.
Cambridge University Press
1981

"No, this is different," he told her. "I have never seen the kind we used ... in a book, anyway. I don't think it's been formally identified."

Mary looked thoughtful. "Well, that's certainly possible. The diversity of fungi is so great that probably only a small fraction of the total number of species on earth have been studied. New types are being discovered all the time. Almost every day, in fact."

He rifled through the book with great interest. At a glance it was or purported to be a comprehensive study of *Amanita muscaria*, the so-called sacred mushroom. One section was devoted to its physical characteristics, morphology, habitat and range, which apparently was worldwide.

Another section included a twenty-odd page discussion of the various chemical agents believed to be responsible for the mushroom's manifold effects. Apparently there was more than one active drug. The chapter intrigued him. But it would have to wait because, suddenly, he wanted to devour the book whole, in its entirety. He quickly skimmed front to back. When he reached the index he returned to the introduction, where on page one he read:

The evidence presented in this study will show that this remarkable mushroom was known from antiquity. The species has a unique morphology and a wide range, though it appears to be confined to the northern hemisphere. It is distinguished by profound psychotropic effects, for which it was indubitably famous in the ancient world. At least one active agent is known to possess powerful aphrodisiacal properties. These appear to depend on dosage....The mushroom is believed to have played an important, perhaps even a central, role in certain pagan fertility rites that were once commonplace throughout the Mediterranean region. The species was known from the Levant to Italy, including parts of North Africa (though this is less well documented), and west as far as northern Spain. It may even have been known as far-East as India...

The sacred mushroom was associated with the cult of the Great Goddess, the Magna Mater.... In the West perhaps the most vital center of the cult was Greece, in both classical and pre-classical periods.... But Greek interest in psychoactive compounds was not limited to fungi. As we will show, a number of other hallucinogenic compounds were also in use... Mushrooms were regarded as sacred to the god Dionysius who, according to Greek tradition, was the divine offspring of Persephone and Zeus.... Dionysius was a complex deity and was often regarded as a form of Zeus himself. The traditional lore held that mushrooms appeared after lightning storms. Hence, the common view that mushrooms were divine, a manifestation of God, the thunderbolt being the principal vehicle of Zeus' divine provenance and power. So it should not be surprising that hallucinogenic mushrooms were sometimes referred to as "the divine ambrosia," or as "the food or nectar of the gods"...

...Dionysius was usually portrayed as androgynous. Among the god's many symbols was the erect phallus, a word that derives from the Greek phallos (pl: phalloi), which was metaphorically synonymous with mykes, the Greek word for mushroom. A poetic equivalent in English might refer to a man's staff of life, his metaphorical rod, or button.... From the same root is derived the modern word for the study of mushrooms: Mycology.... Recently, the classicist scholar Carl Ruck made a convincing case for other fascinating connections. A related word mykene, meaning "the bride of the mushroom," happens to be the root of the word Mycenaea, one of the most prominent of early Greek city-states. It is no minor connection, because in a more general sense the word Mycenaean also refers to the whole of proto-Greek civilization, which flourished during the second millennium BC.

Here Tom deferred to a footnote for more details, then returned to the text:

The root mykene suggests a more important historical role for psychedelic mushrooms than modern scholarship has heretofore acknowledged. This interpretation is consistent with legends known to have been widespread throughout Greece from antiquity. According to these old tales, which date to the earliest known period of Greek culture, each of the Greek city-states was founded by a heroic invader after a sacred ancestral marriage involving ecstatic use of aphrodisiacal fungi....

At this point, the text was augmented with several full-page plates. One was a black and white photograph of a Greek ceremonial vase described by an explanatory note as "a pastoral scene showing temple hierophants gathering *phalloi,* penis-shaped young mushrooms." A second plate was a marble stele from the fifth century BC. It showed two priestesses, or possibly two goddesses. They were serenely facing one another, each with a mushroom in hand. Tom flipped ahead.

The author will build on the excellent work of previous investigators, primarily the ethnomycologist Gordon Wasson, the pioneering chemist Albert Hofman, and the classicist Carl Ruck, in an attempt to show that the species figured prominently in annual pagan rites associated with the famed temple at Eleusis, sacred to the goddess-pair Demeter-Persephone (Kore)....

In classical times, the Greek village of Eleusis, not far from present-day Athens, achieved widespread fame as the home of the legendary Eleusinian mysteries. Annual pagan rites were celebrated at Eleusis without interruption for nearly 2000 years, until the fourth century of the present era.... The mystery rites sacred to Demeter and her daughter Persephone were comprised of two parts, the Lesser and the Greater Mysteries. Each was annually reenacted at a different time of the year, and at a different site. The Lesser Mysteries were celebrated at Agrai in Anthesterion (February), at the start of the Mediterranean spring. These lesser rites were in observance of the legendary abduction of Persephone, said to have occurred at Nysa, a site sacred to Dionysius, whose name means god of Nysa...

Now, he skipped randomly through the book:

...A considerable body of evidence suggests that the sacred mushroom was administered to participants in these Lesser Mysteries...

The Greater Mysteries, on the other hand, were enacted later in the year, at Boedromion (September-October), when Greek farmers traditionally make preparation for the fall planting of cereal grains. This portion of the mysteries was reenacted at Eleusis....Wasson et al. recently presented a convincing case that the Greater Mysteries involved the ceremonial use of a different mushroom, a botanical psychedelic native to the region, namely, the ergot of barley, whose active compound has been shown to be related to lysergic acid dicthylamide...

He looked up. “So, you know about psychedelic mushrooms.”

“I will give that a qualified yes,” Mary said. “Several years ago, I conducted my own investigation of *Amanita muscaria* out of, shall we say, disinterested curiosity. We have them here on the ranch, you know."

He was surprised. "Oh really?"

"Yes, out in the woods behind the barn. The species is not uncommon. They produce fruiting bodies in the early summer, every few years. But not this year, which is curious given what you told me. My curiosity grew out of my scholarly interest in the ancient world. Did Tallie mention that I minored in Middle East Studies?"

"No."

"At a certain point in my research I realized that hallucinogenic mushrooms and other psychoactive compounds were much more important to the ancients than our "modern" scholars have been willing to admit; probably because in our world the social ramifications are almost universally frowned upon. The mystery rites at Eleusis were the best known, but by no means were they the only pagan rites. There was a similar tradition on the island of Samothrace in honor of the mysterious Kuretes. And there were numerous cults devoted to Dionysius, which apparently were very widespread. The most important were the Orphic mysteries, which had a deep influence on Plato, Pythagoras, and even the early Christians. The Phrygian Mysteries were yet another set of observances, celebrated in Asia Minor, present-day Turkey. My guess is that they were similar to the rites at Eleusis. The cult of Osiris was the most ancient of all. As for its origins, we can only guess. They are murky. No doubt, there were many other pagan mystery traditions about which we know absolutely nothing. Our knowledge of the ancient world is, at best, fragmentary..."

As he listened Tom recalled what Tallie had told him.

Mary had paused. "You'll have to excuse me, Tom," she said. "I'm probably boring you to tears. Once I get started on the ancient world I can go on for hours. I find it all so fascinating. But you..."

"Oh no, no. You are not boring," he told her. "Anything but. So, what happened? Why do we know so little?"

She looked relieved. Smiling graciously, she picked up the thread. "Sad to say, Christianity bears a large measure of responsibility. In the fourth century AD, after the emperor Constantine made Christianity the official religion of the Roman Empire, the Church began to suppress the pagan cults."

"But why would the Church or anyone care?"

"Simple. The pagan cults were an embarrassment. You see, despite the spread and dominance of Christianity, the Eleusinian Mysteries had continued to be quite popular. Christian scholars don't like to admit it but the pagan mystery traditions were more than holding their own in the fourth century. Some were, at any rate. It's surely why they were banned. And there is another factor..."

"Another factor?"

"Yes, there is another reason why so little about the mystery cults has come down to us. You see, the initiates into the mysteries were always sworn to secrecy, and the penalty for talking was apparently death. Judging from the rarity of historical accounts, the secrecy oaths were extremely effective, much to the detriment of the historical record. Take the Eleusinian mysteries. They were celebrated annually for nearly 2000 years, a very long time. It's hard for us moderns to appreciate a span of time that long."

Tom was thinking aloud. "That's the equivalent of the entire historical period, since Christ."

"Yes, and during that long epoch of history there were no leaks, with a single possible exception. Tens, perhaps hundreds of millions of initiates took the secrets with them to the grave. It has only been very recently, thanks to the extraordinary work of some able scholars that we have finally gained some insight into what the pagan pageants were really about. In the process, we've also learned about their use of psychoactive compounds. Many find this shocking. Of course, no one really knows if mushrooms were also used in the many other mystery cults of the day. To date, scholarship has mainly focused on the Eleusinian mysteries, but my guess is they were. I suspect that Eleusis was representative, not the exception." She paused for a moment. "I was very surprised when I began to study ancient Greece, at just how extensive the use of psychoactive substances actually was. Did you know, there is evidence that the Greeks bolstered ordinary table wine with powerful drugs?"

"I had no idea..."

"The potency of Greek wine was legendary. There are references to this in the classical literature, in Homer, for example. Greek table wine was so potent it was customary to dilute the wine before serving it to dinner guests. Dilution was necessary, precautionary, because Greek wine was a witches' brew of mind-altering substances. Dilution insured that the desired level of inebriation would be achieved, but no more. In those days, alcohol was not the only, perhaps not even the primary, agent in wine. Today, we bolster wine by adding extra-distilled alcohol. But the practice was unknown in the ancient world."

"Unknown? Why?"

"The distillation process had not yet been discovered; and for this reason the alcoholic content of Greek wine could not have exceeded about fourteen percent. At that point, you see, the alcohol from natural fermentation becomes fatal to the active yeasts that drive the fermentation process. Fourteen percent alcohol is not enough, not nearly enough, to account for the notorious potency of Greek wine. And if the potency was not due to alcohol, well, then other compounds had to be present. It is interesting that the classical Greek language made no distinction between inebriation, ecstasy, and madness. All three states were regarded as increasing stages of the same process of mind-alteration. Still, and this is important, two very different classes of psychoactive drugs are known to have been in use; because, you see, the ancient Greeks maintained a firm distinction between the sacred and the profane. Hallucinogenic mushrooms were in the realm of the divine, hence, were a part of pagan religious life. Alcoholic brews, on the other hand, were a part of the secular world and were never involved in the mystery rites. On the contrary, initiates at Eleusis were actually required to purify themselves by abstaining from wine for several days before participating. The ancients believed that the realms of the sacred and the profane were incompatible. Like oil and water, the two did not mix."

"You mentioned an exception."

She laughed. "Oh, that. Yes, well, it seems that in the year 415 BC a series of profanations occurred. Apparently, the 'jet set' crowd in Athens managed to acquire some of the hallucinogenic compounds used at Eleusis. In the rites the mushrooms were usually taken in a liquid form, the famous *kykeon.* The result was a major scandal. The Athenian swingers began using the stuff for their own personal entertainment. Apparently there were wild and debauched parties, and when this became known it triggered a national scandal. The authorities cracked down and meted out harsh penalties. The Greeks did not look kindly upon profanations of sacred ceremonial rites. They made a stern example of the offenders. You will find this covered in O'Sullivan's book. Contrary to our uninformed modern perception, the ancients were neither stupid nor ignorant. Sacred substances were not forbidden. But their use was strictly regulated, and for good reason."

"Why regulated?"

"To protect the family and society. In those days hallucinogenic substances could be legally gathered and administered only by temple priests or priestesses; and the hierophants had to undergo years of training and preparation. They were also bound by solemn vows. Their lives were dedicated to the service of the goddess. The male priests were often eunuchs, just to give you some idea of the level of commitment. We only know about the scandal thanks to several references in classical literature. Plutarch mentions it. Also, we have a fragment of a comedy by the playwright Eupolis written shortly after the scandal. In one key scene an informant gives testimony in a courtroom. The witness explains to the judge how he knew for certain that the accused had in fact been using the illicit substance. Evidently, the accused man had attempted to bribe the witness to say that he had only eaten porridge. But the informer knew it was a lie because the man had purple barley groats on his mustache. At Eleusis, you see, the potion's active ingredient was the extract of an ergot, a kind of fungus that grows on barley and has a well-known characteristic: it stains purple. In the comedy the punch line involved a play on words. In classical Greek 'crumbs of barley' could also mean 'purples of barley.' No doubt, the line produced howls of laughter among the initiates in the audience, to anyone who was in-the-know."

From Mary's tone it was apparent that she had little enthusiasm for the mushroom, whatever its historical significance.

"And you are satisfied with O'Sullivan's account?"

"Oh yes. His book is a most impressive piece of research. O'Sullivan is the best type of scholar. By that I mean he is relentless. He pursues the trail of evidence through a mountain of classical material, indeed, with all the rigor of an archaeologist excavating a tel. What more can one ask of a scholar? Not many of them have that kind of courage. It takes nerve to follow the evidence no matter where it leads. O'Sullivan's near unfailing intuition is also impressive. He follows his proverbial nose. To be sure, he does rely heavily on previous detective work by several other no less competent scholars, men such as Gordon Wasson, Albert Hofman, and Carl A.P. Ruck. In case you don't know,

Wasson and his wife were the innovators who created the new field of ethnomycology; based on their researches into the psychoactive mushrooms of Mexico and Central America. Hofman was the pioneering chemist who in 1943 accidentally discovered lysergic acid, LSD; although it is more accurate to say he rediscovered it. And Ruck was a very able classicist. O'Sullivan built on the work of these capable individuals, and so, was able to confirm the role of the mushroom in the lesser mysteries at Eleusis. And, well, I think that just about covers it."

"I gather you are not particularly enamored of the whole business."

"You mean the mushroom?"

"Yes."

"Perfectly right. I am not 'enamored,' as you say. If you ask me..."

He cut her off with a missile. "But why condemn something you haven't tried? I mean, if you haven't done it yourself...how could you possibly know?"

Her benevolent smile never wavered. "Tom, Tom," she said, "You are jumping to conclusions. I am more sympathetic to your point of view than you seem to think."

"Huh?"

"Who said I didn't try it? I never said that."

"But you said..."

"No, on the contrary..."

"You mean..."

"I am not uninitiated. Several years ago I did use the sacred mushroom. The *muscaria* variety. Indeed, on several occasions."

"You did? No kidding?"

She nodded.

"Well?"

"Well what?"

"So what did you think? Were you blown away?"

"I've told you, it was a case of honest curiosity."

"But..."

"Allow me to finish. I make it a habit, at least I hope I do, not to speak from ignorance. Narrowly opinionated people drive me to distraction, especially the Christian moralizers. They're the absolute worst. That's how my father was, you see. He was the stereotypic stiff-necked fire-breathing Christian minister of the Baptist persuasion. It probably explains why I cannot abide missionaries. I suppose it's also why I go to such lengths in the other direction, a compensatory habit of mine. As I said, we've seen the fruiting bodies, many times, here on the ranch. The species is not uncommon. I may be forty-seven, but I do speak from personal knowledge. Direct experience. The mushroom high was, how shall I say, quite remarkable. Yes. In its...amazing way. Remarkable. I have no regrets about using it. Understand, I am not opposed in principle to 'getting high.' I happen to believe that the human drive to experience altered and higher states of consciousness is a natural and healthy part of our nature. But I would go even farther. I suspect that "getting high" is an essential part of the human condition. Rooted in history and human consciousness. It's a human need, and may even have a biological basis. Something genetic, built into us. No, my reservations are of an entirely different sort."

"Your reservations?"

"Certainly."

"Of a different sort?"

Mary reflected for a moment. "You've heard of Richard Alpert?"

"Ram dass."

"Yes. Formerly, a member of the psychology department at Harvard, where he was an associate of Timothy Leary, the so-called LSD guru."

"It was Leary who said: 'Turn on, tune in, drop out...'"

"Yes, a somewhat regrettable phrase. Well, in one of his books, Alpert, I prefer to use his original name, relates an interesting story, one relevant to your question. It seems that soon after he departed the Harvard LSD scene, Alpert traveled to India. At that time in his life he was a self-described seeker after wisdom, a hippie in search of the Higher Self. Enlightenment. Call it what you will. Well, it seems that while in India, by luck or chance, Alpert happened to meet an authentic holy man, a Hindu saint, one of the numberless adepts who, though they remain almost unknown in the West, have for millennia made India the spiritual powerhouse of the planet. In his book Alpert describes the meeting with this great sage and how it changed his life. The way he tells it, the change was permanent, which to my way of thinking is the acid test."

Mary covered her mouth to stifle a laugh.

Tom chuckled with her.

"Oh dear. Please pardon the pun. In any event, Alpert claims this saintly man read his most private thoughts, told him intimate things about his own mother, things no stranger could possibly have known. Well, as you might expect, this got Alpert's immediate and undivided attention. Soon after, the saint asked him for 'the medicine.' At first, Alpert did not understand what the yogi was talking about. But, eventually it dawned on him that the man wanted the LSD. At the time, Alpert was carrying on his person a major cache of hallucinogens, LSD of great purity..."

"Maybe windowpane."

"Possibly. I have no idea. The point is that when he produced 'the medicine' the saint promptly downed the entire vial. Enough LSD to unhinge a pachyderm."

"What happened?"

"That's precisely the point. Nothing happened."

"Nothing?"

"That's the way Alpert tells it. Absolutely nothing. And I see no reason to question the veracity of his account."

"But..."

"Zero. Nought. According to Alpert the saint was totally unfazed by the most massive dose. And, unless I am mistaken, this was the intended message. And, to my way of thinking, that pretty much disposes of your question about the mushroom. Of course, I grant you, it is possible that Alpert fabricated the whole story. Made it up. It's even possible that his brain had been addled by the drug. He might have confused fantasy with reality. Possible, but in my view not likely. No, I tend to believe him. I'm willing to give him the benefit of the doubt. You see, we are free beings, burdened, so to speak, with free will. The point, Tom, is that one can choose to remain a bullfrog croaking in a puddle. Or..." She paused.

"Or?"

"Or, one can reckon with the sea..."

The words rocked him. Suddenly, he was listening to his own windy thoughts through the clapboards. For a brief moment his thoughts seemed like ripples on a vast millpond. When he looked up she was smiling mysteriously.

FORTY THREE

For a dozen miles he played peek-a-boo with the Never Summer Range. The storm had spent its fury but the weather would not make up its mind. The sky above the mountains was a mixing bowl, dark and light, a patchwork of clouds tumbling sideways in a hurry.

His head was awash. The events at Bowen Gulch were too fresh to have any real perspective, yet.

The highway north to Willow Creek Pass was like a tunnel. The high country was never far away, only a few miles east of the road. But the highway followed the snaking stream, lined with willows; hence the name, and the nearest ridges blocked his field of view. The mountains were mostly hidden. Occasionally though as he rounded a turn, he caught a glimpse of one of the high peaks dusted with new snow. Each glimpse was a thrill.

It occurred to him that he lived for such moments, that most of his young life had been spent anticipating, tantalized by the next peak. Another pun.

They don't call them peaks for nothing.

This rekindled the philosopher in him, and he took up the matter. Were not "waiting" and "glimpsing" closely related, two different parts of the same picture, like the "figure" and "ground" of a *gestalt?* Yes, but which was the figure, and which the ground? No need to reason it out. He knew the answer. All of his life, even before taking up philosophy at the university, he had been a hard-core glimpse addict. Was not the glimpse the thing? Yes, the meaning was in the glimpse. The peak. Still, he found the answer dissatisfying, somehow cold and disturbing; for he could not shake what this implied, the plain and undeniable fact that the average human spends most of his or her life in meaningless tedium, awaiting the next fleeting moment of joy and happiness. Years of waiting for a few seconds of pleasure, a glimpse of happiness. It did not seem like near enough. He was still thinking about it when Park View Mountain came into view, capped by fresh snow. At the sight he tabled the silly dialogue in his head.

Rubbish.

She had finally opened to him that day in the cafe, over lunch, told him about Luther and the rest of the story about San Francisco; how during her months there she had consulted with specialists at the University of California Medical Center about the migraines. The doctors ordered a battery of medical tests. But it was the detailed history they took that led to the breakthrough. They discovered that her migraines were always preceded by what they called a "prodromal sign," in her case flashes of red light that appeared in her peripheral vision. She found that if she caught this in time she could head off an attack by using a certain relaxation exercise that she learned from one of the doctors. Suddenly, she was holding her own, keeping the darkness at bay. She had not suffered a full-blown migraine, since. Although the chronic pain continued she found it was manageable. She could deal with it.

The new measure of control that she regained over her life was revelatory, because the issue of control, or rather its lack, had always been her main trigger. The issue was closely linked to her difficulties with men, and no doubt traced to the anorexia of her youth. For the first time in years, life was full of hope. She looked forward to getting up each morning.

The breakthrough had given her a new lease, that is, until further testing revealed a tiny aneurysm deep in her brain. The specialist told her that a surgical repair was not possible due to the location. It was inoperable, a death sentence. There was no clear or obvious link to the migraines.

"I'm sorry, but there is nothing more we can do for you," the specialist told her. He also declined to give an estimate about how long she could expect to live. Weeks? Years? Months? "There's no way to tell," he told her. "It would only be a guess." He tried to reassure her. "The best advice I can give you is to try to forget about the aneurysm and just move on with your life. You might live for years."

It was why she had not written sooner. What was the point? Each day was like Russian roulette, a roll of the die. She did not know when she got up in the morning whether or not she would live to see the sun set. It seemed pointless to look ahead, or to make plans; and there was another factor. She felt it would not be fair to involve Tom in her tragic life. She did not want to hurt him, any more than she already had.

"I should have told you straight out, first thing."

"Yes, but that doesn't change how I feel."

As time passed her views had evolved. Eventually she reached a fatal acceptance of her likely imminent demise and decided that she must not allow this to prevent her from living fully in whatever time she had left. She would not retreat from life she would go for it.

"I'm living on borrowed time. I can drop dead at any moment."

"Then we'll have to make every second count, won't we."

He was listening to Mary in the kitchen when he realized from something she said that she did not know. For whatever reason Tallie had not yet told her.

They heard her moving around upstairs. "She's up."

He went up to tell her the meal was ready, and found her singing in the shower in a foreign language. It sounded French but he was not sure. He pulled back a corner of the shower curtain and copped a peek. She was enveloped in steam, her head covered with soap and suds. She giggled at him. The girl had no shame.

"Is that French?"

"Oui" she said, as a rivulet streamed off the end of her nose. More giggling. She yanked the drape closed. "Hey, you're letting the cold in!"

What a loopy girl.

"We'll eat when you're ready."

OK, she was fluent in French. Now as he thought about it he recalled that in Florida she had been reading *Madame Bovary* in French. As he went back downstairs he wondered how many other languages she knew.

"Oh, it runs in the family," Mary told him. "Her father was a linguist. Tallie picks up languages with the same ease that she learned to ride. She speaks Italian, German, and Spanish that I know of. I'm not sure what else."

Mary had set a plate before him loaded with turkey and other good things when Tallie glided in, barefoot as usual. Before he could take a bite she planted herself in his lap and wrapped her thin arms around his neck. She was freshly scrubbed, her hair still wet, fragrant with soap and shampoo. She laid her forehead against his and looked in his eyes.

"I'm proud of you, Tom Lacey."

"Why?"

"For what you did."

"What did I do?"

"The right thing. I knew you would."

Mary was by the stove. "We are both proud of you, Tom," she said.

He stroked her cheek. "Are you going to join us?

Mary laughed. "Hey, the turkey was *her* idea. She'll probably eat enough to feed a battalion."

"I wove cranberry," she said, playing with the words, her eyes sparkling, but when he looked closer he saw something else, something that unnerved him: exhaustion.

Now, he crested the pass. Before starting down the north side he pulled off at an overlook and cut the engine. The site commanded a sweeping vista of North Park. The 8,000-foot high valley sprawled below for many miles, north to the Wyoming border. In the Northwest, beyond the sage flatlands, rose the distant ramparts of the Park Range. But he hardly noticed. He was staring vacant-eyed through the windshield. He had already wept himself dry; he had no more tears. In a million years he could not have described how he felt in that moment, inert somehow, stripped of substance and emotion, even sadness. He was in a null place, a dead zone, hollow, empty as the valley below. His only feeling was an overwhelming sense of loss. He knew he was on the brink.

What was life but a labyrinth of fucked up days and cruel nights, conceived for some unknowable purpose? Why are we here? To endure one slap in the face after another, before passing out of existence, back wherever it was we came from, the far country.

Yes, life was a death sentence, the butt of some twisted joke. One great cosmic rebuke. He lifted his eyes to the gray-scrabble sky. Out of the emptiness came a movement or maybe a murmur. He was at a loss to know which, or to identify the source. Somehow though he sensed it was not outside, not out there. It seemed to be moving through him. Suddenly he knew he was listening to the throbbing of his own heart, hearing his pulse in one ear.

He felt tremendous heat. Suddenly he was burning up. On impulse he grabbed a pen, then, ransacked the front seat and the glove box for a scrap of paper, something, anything to write on. He found an envelope and scribbled a few lines on the back. He worked in a fever. Words, no, phrases, fully formed ideas, images, spilled out of him. When he had used up one side he continued on the back in the same fashion. He rearranged a few lines. Somehow he knew what changes to make. He completed a last draft. He could scarcely believe what he had written:

The pleasure of your peerless company
Was everything in which my heart delights.
Your lilting voice like some epiphany,

The promise of a hundred carefree nights.
Each playful syllable a fond caress.
Each word a song, and in your childlike charms
a world of feeling I could not express
until I wrapped you in my aching arms.
How sprightly you appeared upon the deck.
Your impish smile, your knowing look a dare.
I coaxed you gently once upon the neck,
your slender fingers in my wavy hair.
My hand upon your flushed and heaving breast.
Another urgent look, and we undressed.

FORTY FOUR

Olsen's tiny camper was parked on the edge of the landing, as before, surrounded by the usual flotsam and jetsam of a small post & pole operation: hand tools, drums of oil, one with a handle pump, five-gallon gas cans, a roll of steel cable, various lengths of chain, grappling irons, chain binders arrayed in a neat row, a cutting torch with gas and oxygen tanks, new and old tires, a pile of odd-shaped pieces of iron, miscellaneous gear...

The dock though was strangely empty, vacant of product. This late in the season the landing should have been piled high with fence posts, sorted in different sizes and lengths. There also should have been a separate mountain of corral poles. Olsen's flatbed was parked beside the antiquated crane that he had picked up for next to nothing. The year before, the old man had rebuilt its diesel engine at his yard and machine shop on the outskirts of Walden. Olsen had re-engineered the crane, customizing it to suit the small scale of his operation.

But where was the old man? The tiny trailer was empty. The woods were silent. Yet, his pickup was parked nearby.

Carl has to be around someplace.

Tom walked aimlessly about the yard stretching his legs. He kicked a stone and whistled a tune. After a few minutes he returned to the Toyota and laid on the horn several times to announce his arrival. He switched on the radio and waited in the cab with the door open for the old man to return. A few minutes later Carl emerged from the trees behind the camper carrying a roll of toilet paper. The first thing Tom noticed was the paunch, or rather, its absence. Olsen was visibly thinner.

Not good.

It looked as though Carl had endured another operation. Tom drew some consolation from the fact that the old man moved with the same graceful step, like a cat. Olsen was as nimble as ever. Sympathy was out of the question.

Best show no concern.

Over the last few miles as he descended the north side of the pass and followed the access road up Snyder Creek, he had rehearsed how to dish up old "Battery Acid" a dose of his own medicine. Now, he meant to deliver.

"Afternoon, Carl," he said with a sweep of his eye across the near-empty landing. "I see you been terribly busy."

The old man grinned but he wasn't buying. "I didn't think you'd be back."

"I wasn't sure myself. But here I am."

The old man rubbed his hand across a scratchy two-day beard. He tipped his sweat-stained fedora up in front, exposing his pale scalp which was nearly bald. The exposed flesh was white as snow, in contrast with his red face, fired by sun. His rough hands were akimbo, resting on his hips. "I wouldn't a bet a wooden nickel on you coming back. You were so hot and heavy into the big stuff."

"Come on, Carl. I wasn't that heavy into it."

"The hell you weren't," Olsen said. "You was shit on wheels to knock down everything this side of Chicago. So, what made you up and quit? Or did they fire your ass?"

"Nobody fired me," he shot back. Tom did not want to be having this conversation. He had not been here five minutes, and already he was fuming at Olsen. He was even more irritated with himself. He took evasive action. "I don't see much in the way of production," he said, motioning toward the empty landing. "What's the holdup?" But there was more edge in his voice than he meant to deliver.

Olsen was stone-faced. "You're a funny guy, you know that?"

"D'you want me back, or not?"

Carl lifted a hand, as if to placate him. "Now don't get all peeved. I was only being curious. Sure I want you back. You bet."

"I could start tomorrow."

Olsen nodded. "OK. That's soon enough."

"If you don't have any objections, I'll camp where I did before."

"Camp where you please."

"Alright. I will."

There was a strained silence. But Carl's craggy face slowly morphed into a grin. "For your information, smart ass, the landing's near empty 'cause I 'jes sent out three big loads, day before yestidee. I had another cutter, guy was supposed to be here early this morning. But he's a no show and...oh hey, your old pal Roper's back. Drove in from San Francisca coupla' weeks ago."

"Rope a dope!"

"Late, as usual."

"That's great. I haven't seen him since last year."

"Don't know if you heard, but after you and him finished last September, he went a gallivantin' off to wild and wonderful California. Unless I'm mistaken, that was the last anybody around here saw of him. I only got a couple of postcards, and a short letter about six weeks ago. Roper didn't like California."

"Did he say why?"

"Ask him yourself. He's working up the road a piece." The old man chuckled. "Brought his new girl friend with him. Name of Rebecca."

"Yeah?"

Olsen winked. "She's a live wire."

Tom climbed back in his rig and motored up the road above the landing. A half-mile on, he found Roper sitting on a log by the side of the road, elbows on his knees. The large man was having a smoke. He had wood chips on his clothes and in his hair. A chainsaw lay beside his feet.

"Make my day," Roper said.

"Good to see you, Dave."

The man blew smoke. "Where *you* been?"

"You wouldn't believe it."

"Try me."

"I will. But first I need to unload."

"Come by for dinner. Same as before."

"I heard about Rebecca."

Roper rolled his eyes.

Tom reestablished his old camp on the same point of land where he had summered the year before. Why move? The site was unsurpassed. It lay on elevated ground in a stand of mature lodgepoles, at the confluence of Snyder Creek and a tributary branch. Firewood was abundant and close at hand. The site had plenty of cover and easy proximity to stream-side meadows, and superlative views in three directions; and it was optimal for another reason. The open aspect made the most of the slightest breeze, which helped to keep down the mosquitoes. A fast-flowing cold spring along the edge of the meadow about fifty yards upstream, supplied drinking water of unexcelled purity. Water suitable for washing was available by the bucket from a beaver pond closer still.

Both branches were a succession of beaver ponds backed up behind well-maintained dams. The pools were full of native brookies. As Tom unloaded he watched them hit the surface.

Shortly before sundown he showed up at Roper's camp. His old friend had graduated from living in a tent and was now residing in the lap of luxury, a 25-foot Airstream camper. A woman was feeding a dog as he drove up. She was about six feet tall, had short blonde hair, and was very good looking.

"Are you Tom.?

"Yes. Nice to meet you."

"I'm Rebecca. This here's Shep." She motioned to the dog. "Dave's expecting you. Go on in. Dinner's on the stove."

Roper was standing by the table lighting a kerosene lantern. After he blew out the match, he carefully inserted the glass chimney back in the four-pronged holder. He adjusted the wick and slid the lantern to the middle of the table. Then he moved to the stove and lifted the top from a great steaming cast iron kettle and began serving up a stew of some kind.

"Goulash," Roper said with relish, and handed Tom a large bowl full – big chunks of meat swimming in a sea of gravy. "Venison stew. Wait 'til you taste it. This morning, the old man cut me some juicy steaks off that big buck he shot last week. The first of the season. Oh, sorry, I forgot what a finicky eater you are."

"Tom doesn't eat venison?" Rebecca said. She had followed him in.

"Well, I..." But Roper cut him off.

"Why do you think he's such a fly-weight? From subsisting on rabbit food." The tone was mildly derogatory.

The truth was, Tom did eat meat on occasion, though he eschewed the red variety.

"It's what there is. He'll eat it or starve."

"I'll manage," said Tom.

"Oh dear, if I had known," Rebecca offered, trying to be helpful. "There's taters and veggies in there, too. You can eat around the meat."

He made out OK though. In fact, the stew was delicious.

It turned out that Rebecca was a radio talk-show hostess, and had her own midday show on a San Francisco station. Anyway, she did before she ran off with Roper.

"We met in the tenderloin," she said. "But I'm originally from Tucson." She gave Roper a meaningful look that Tom was at a loss to cipher, some kind of private language. But apparently Roper failed to respond because she then gave him a vicious kick under the table. The big man took it without a flinch and went on eating as if nothing had happened. He had already loaded up on seconds.

"If he thinks I'm going to be his squaw out here in the wilderness, and cook and clean for him…" Evidently too pissed to finish, she got up and began clearing the table. The moment she did, Roper smiled. He was down to wiping the bowl with his bread.

"Carl's got 200 acres of dog hair," Dave said. "Enough pole timber to keep us busy 'til the snow flies."

"Looks like he had another operation."

"Yes. Last April. They removed his appendix, among other things. He's slowed down some."

"I noticed."

"Doesn't do as much hauling as he used to. Another guy does a lot of it now. The other day I was kidding Carl about it. He showed me his new pink scar, across his lower gut. Almost with pride, like 'Look at what I survived..' I said, Carl, you'll probably outlive all of us. I was only trying to commiserate but know what he said?"

"I can guess."

"'I aim to,' he said. He wasn't kidding either. He meant it, the old fart."

"He thinks he'll last forever."

"So whatcha' been up to?"

He told them about Bobby Lighthorse, the mushroom and the wild goings on. He was not far into it when he saw Roper looking at him in disbelief. Roper's face was stern but his eyes were neutral, maybe even smiling. "Lacey, you're bat crazy. You know that?"

"Tell me something I *don't* know."

Rebecca had been busy in the kitchen and now served up desert, pineapple upside-down cake which Tom thought was appropriate under the circumstances. It was melt-in-your mouth good. When they finished she cozied up beside Roper. The two kissed lightly on the lips. His cue that it was time to leave.

He was going out the door when Roper said, "Let's hit those high lakes, bud."

"Sounds like a plan."

"Next weekend."

"OK."

In subsequent days, the two fell into an easy work routine that never varied much. Sweat and wood chips became the order of things, bracketed by frequent good times. Tom usually finished work by 2 P.M., after which, he would clean up, then fish for his supper, or maybe go for a hike along Snyder Creek. Broad meadows up and down the stream were thick with willows, except where the beavers had thinned them out. There were also scattered stands of quaking aspen.

High summer had come to the Rockies. The annual wild flower pageant was underway. The meadows were glorious, a mosaic of color. Some days he went upstream, other days down. The direction hardly mattered. Either way afforded equal opportunities to fish and view wildlife, and equally stunning vistas of dusk gathering over the Never Summer Range. The continental divide was only three miles upstream.

Carl was visibly weaker and had lost a step. However, the old man was ornery as ever; he was hard to kill. After the operation his family had encouraged him to retire to an easy chair and be fussed over by daughters and grandkids. But Carl would have none of it. Long before, he had charted his path and he would stay the course. In Carl's case it was the path of greatest resistance, forged out of sheer defiance. The old buzzard would work until he dropped.

Though he had slowed, Carl remained as inventive as ever. He had bolted a small boom to the back of an ancient stripped-down jeep, which he also outfitted with a winch and cable. He was now using the custom-made rig to skid the poles and posts out of the woods. Not only did he handle all of the skidding, loading, and mechanical repairs himself, about every other week, when he felt up to it, Carl made a delivery to Denver or Greeley, and occasionally to points beyond. Somehow he got as much work done as most men half his age. This was due in about equal measure to his tenacity, his general cussedness, and to the fact that he knew how to conserve his energy. Carl never hurried. He had sworn off haste because, as he put it: "Ever' time I hurry things go straight to the devil." He worked in short and measured bursts of activity, between frequent food breaks and rest stops; and because he started early and was content to finish late, most days he managed to get in a full day's work. Sometimes, to stay caught up, he even worked on Saturday; in other words, a six-day week; not bad for a tumorous seventy-eight year-old great-granddad.

For all of that, there was no perceptible strain. The old man knew his limits. Though Olsen was as cantankerous as ever, there was no pressure to "get out the cut." The atmosphere around the landing was relaxed and cordial. Carl set a modest weekly quota. He was happy to get out one load a week, and it was easily met. Production happened at its own speed. In the boss's unhurried view of things, that was "damn soon enough."

Dave Roper, who was a minister's son, was circumspect when Tom tried to engage him in philosophy. "Ask me no questions and I'll tell you no lies" was his catch-all comment about the human condition. Tom got the message and wisely refrained; but he often kidded his friend about his giantism. Roper, who was six-feet-seven, faced a multitude of challenges in a world designed for shorter men. Doorways were a constant challenge. Much of the time he had to lower his head to gain passage to wherever, and he often bumped it anyway. Beds were another issue as no ordinary mattress could accommodate him. The big man had grown accustomed to sleeping with his feet dangling in space; and he often suffered from cold toes. Showering posed another kind of challenge. Roper had to stoop just to get wet in his camper's tiny shower stall.

All the same, he never complained. His standard rejoinder was "Hey, it's all relative." When pressed, Roper would remind him "being an extra-large does have certain compensatory advantages, especially with regard to the ladies." Roper's relative advantage was eleven inches long.

He and Rebecca were fortunate in that the Airstream had a propane powered water heater. It worked on-demand and afforded the luxury of hot water in their small kitchen, and hot showers. But the built-in storage tank held only about 30 gallons. So, every four or five days Roper had to motor down to the cold water spring near Tom's camp to fill up a great cistern in the back of his pickup, using a hand pump. Which he would then download into the camper's water tank. They used the same spring water for drinking and cooking.

One day, when Roper had come for a refill, Tom fished two cold beers out of the spring and handed one to his friend. As usual, the big man had brought Shep along. The dog was about two-years old and loved to ride in the back of the truck with his nose in the wind. He was of mixed parentage, half Australian Border Collie and half coyote. Technically, he was a mongrel, but Shep was no ordinary dog. Roper claimed that the animal was so intelligent that the only discipline he ever needed was an occasional mild scolding. He had never known a leash or a chain, nor had he ever been confined to a pen, not even in the big city. According to Roper, the San Francisco canine patrol picked him up on one occasion. Roper had been compelled to pay a stiff fine to reclaim him. However, the incident only proved the dog's remarkable capacity for single-trial learning. Thereafter, Shep roamed as freely as ever, in flagrant violation of the city's leash ordinance, and successfully evaded the dogcatcher. He was never apprehended again.

"Did you hear the coyotes, last night?"

"Yeah. They were howling up a storm. And really close."

"They were more than close. They came into my camp to check him out." He indicated Shep.

"Really?"

"That's the second time this week."

"Did they fight? I read somewhere that coyotes gang up on dogs and kill them."

"No. But it was pretty cool."

"So what happened?"

"Tell you what. I'll leave him with you. Maybe tonight you'll get lucky and they'll come back in. You need to see it."

"But Shep is your dog. Will he stay with me?"

"Oh hell yes. Watch this."

When Roper topped off the cistern, he summoned the dog, talked to him, and gave him some friendly strokes. The tail-wagger nuzzled his hand with his nose.

"Sit, Shep," he said. The dog sat on command. "Now, stay, boy. You stay with Tom." That was all. That was it. When Roper left with his load of water Shep remained behind and obediently followed Tom back to his camp.

The last thing Roper said was, "Don't yell at him. Just talk normal."

Later, that evening, when Tom had eaten and was cleaning up by the last light of his fire, he noticed a sudden flash through the timber, a light where no light ought to be. The weather had been overcast for several days.

A wild fire?

Almost immediately, though, he dismissed the notion. He also ruled out lightning. There was no indication of a storm, nor any wildfire. What, then?

He set aside the half-scrubbed cook pot, rinsed his hands, and moved away from the circle of light. He stepped out from under the lodgepoles into the pitch-dark meadow. Gradually, his eyes adjusted.

The evening was brisk. A stiff breeze was in the works. The sky was a scrambled sea of clouds. Shep was beside him now, nose high in the wind. The same strange light briefly reappeared, in the direction of Radial Mountain. It was kind of diffuse, but when the clouds briefly parted the sky yielded up its secret. It was the full moon rising.

Damn.

Now, came an unearthly howl from somewhere behind camp. It was not far off. One lonely yowl followed by many others, a high-pitched chorus of yips.

Shep pricked up his ears. The dog was now agitated and began to speak, not a howl, more like a low moan. His nose tested the air in different directions. Suddenly, Shep disappeared into the dark timber.

Tom followed.

There was no seeing, now. The moon had retreated. Tom had to feel his way in darkness so complete he could barely see the nearest trees. The yowling and yipping intensified until the roiling mists, back-lit more and more brightly by the diffuse light, gradually parted, and the cool and crazy rays of the full moon filtered down, gently illuminating the vertical face of every lodgepole in a faint blue light.

The howling was louder now and very close. Shep seemed drawn to it as if by some ancestral tide. No mistake. It was the call of the wild. Instead of trying to restrain him, Tom spoke soothing words. The pack was now within maybe fifty yards, just out of sight. All at once, the howling stopped. He could hear the troop moving about in the timber. Now, there was a renewed series of gentle yips. They were speaking to one another.

Shep moved cautiously forward. The moon was full up, now, the woods backlit brightly, almost as bright as day. The leader came in closer than the rest of the pack.

There followed an exploratory meeting in the moonlight, nose to nose, a kind of canine reunion as they sniffed each other, but only briefly. After maybe half a minute the ways parted and the true bloods were gone in the night.

Later, the howling resumed from a great distance.

FORTY FIVE

That weekend Tom and Dave fished the high lakes at the head of the Michigan River. The lakes were like a string of pearls, each at a different elevation, all nestled in a spectacular bowl on the east side of Mount Richthofen, one of the highest peaks in the Never Summer Range.

They set out after work Friday afternoon in Tom's rig. Earlier, Rebecca had conscripted Roper's truck and headed off to Denver to visit a girlfriend.

The trailhead was at the end of a passable jeep track, about a mile beyond where the highway over Cameron Pass leaves the valley floor of the Michigan River to begin its steep ascent. By the time they reached the end of the road the sun had already dipped behind the Nokhu Crags, a jumble of pinnacles that dominate the western skyline in that part of the valley. The jagged peaks were punctuated by graceful snowfields.

It was a five-mile hike up the valley to the string of high lakes. Roper had brought Shep along – not that he stayed with them. Most of the time the dog ranged far ahead, checking out everything, though sometimes he also backtracked to the right or left, following his nose through his olfactory world.

They reached the first small lake before dusk, but pushed on past it, climbing steadily through a rocky garden of wild flowers.

They moved up the string until they reached the largest lake. Immediately they dumped their packs, pulled out their rods and began fishing from the shore. Within ten minutes they had their supper, a mess of pan-sized native trout. Roper pulled out a fillet knife and went to work cleaning the fish while Tom stumbled about the gathering dark collecting wood, then built a fire. They cooked them up in butter, with a dash of salt and pepper. They fried potatoes in a second skillet. The fare was simple, nothing fancy, but the taste was out of this world. They washed it down with a bottle of red wine Roper had packed in.

They were in complete darkness by the time they finished eating. The crackling fire was burning down to a red pile of coals. The night was warm and clear, so they decided to forego the tent and sleep out under the stars. They emptied their stuff sacks, plumped up their down bags, and stretched out with their hands beneath their heads under a dome big enough to make any mortal feel small and insignificant.

The thin mountain air was like an open window, into... what? Infinity, maybe. The dark sky was a swathe of glittering diamonds.

"Can you believe it? So many stars..."

"I think I must have died and gone to heaven."

"Yeah."

"It doesn't get any better than *this*."

"Just think of all those poor city folks. In their ticky-tacky homes and cramped apartments. With noisy neighbors and grouchy landlords. People who in all their lives have never seen the sky like this. The way it really and truly is."

"They don't know what they're missing."

"If we could figure out a way to bottle this and market it, hell, we'd be millionaires."

It was true. Tom had never seen the Milky Way so bright, not even the night he did the mushrooms. The aching beauty of it and the vastness canceled their conversation. What was the point? For a long time they studied the heavens, contemplating the wonder of it all. Occasionally, a silent meteor streaked across the vault.

"Roper, what are we doing here?" Tom finally said.

"Fishing."

"Come on. You know what I mean."

"Tom the philosopher."

"I know you're deeper than you let on."

"You think too much."

The only answer was an all-encompassing beauty more profound than Kant, or any book about philosophy.

Next day, they fished all of the pools, up and down. Soon they had plenty and were throwing them back, fishing for the fun of it.

Sometime about noon they built a fire and had their lunch. Afterward, they set out to explore the tundra bowl beyond the lakes. The entire basin was above timberline. A person could walk for miles in just about any direction.

Roper produced a camera from his pack. That was when they noticed a group of large animals grazing on a distant slope, across the basin. Roper identified them through his viewfinder.

"Dahl sheep."

"Big horns!"

"Yeah."

The sheep, about two dozen in number, were contentedly browsing a lush meadow just below Thunder Pass, on the border of Rocky Mountain National Park. The small herd was about a half-mile off.

"This must be the group Carl's been talking about. He says they've never been hunted. They live mostly inside the park."

"Let's see how close we can get."

They set out and within about twenty minutes had covered most of the distance. As they hiked Roper used verbal commands to restrain Shep so the dog would not run ahead and spook the herd. Before they covered the last two hundred yards he ordered the dog to sit and remain behind. On command, Shep sat down and waited, just as he was told.

At the last, the slope was very steep. Tom expected the sheep to bolt at any moment. But they never did. The big-horns had no fear whatsoever. Carl was right. They held their ground. They actually turned from their meal of grass to watch the humans. The sheep seemed genuinely curious. They were shedding the last of their heavy winter coats. Loose tufts of wool hung from their bellies, drifting in the gentle breeze.

They got in very close, within about twenty feet. The killer was that, by then, Roper had shot all of his film.

FORTY SIX

The fishing was so good they stayed an extra night. Sometime Monday after lunch, they started back. Instead of retracing their steps down the valley though they clambered up and over the ridge north of the high lakes, and with Shep in the lead made their way over some rough and trackless back country toward the Nokhu Crags.

They came down about a mile from the crags through an old timber sale. The entire side of the mountain had been ripped apart by tractors, hundreds of acres thrashed beyond belief. They stumbled in shocked silence through a scene of devastation, a ragged stump field as far as they could see. Deep tractor ruts exposed forlorn masses of dead roots like wild beards. Whole stumps had been disemboweled along with their root wads and lay tipped at crazy angles. They climbed over man-made moraines of rocky earth and skirted piles of logs left behind to decay. Water was everywhere and on the move. An icy brook coursed down an anonymous access road through deeply rutted tracks. The road had washed out long ago, another scar from a forgotten past, the work of unknown men who had since moved on. It was mid-afternoon when they made it back to the pickup.

After regaining the highway, Tom pointed his truck south toward Willow Creek Pass. As the miles sped by the subject of Tallie arose again. They had talked about her often, because Tom could not get her off his mind. Roper listened.

"Sounds like you got it bad, man."

"I just can't throw this one back."

"Well, partner, I hope you know what you're doing. I never heard of an anorexic who ever had it together. Or much of anything on the ball for that matter."

"You haven't met her."

"From what you've told me, I'm looking forward to it."

"You'll get your chance. She's coming."

"Yeah? When?"

"Don't know, but soon. I talked with her last weekend."

When their conversation played out Roper picked up a Denver station on the dial. It was Gordon Lightfoot singing *Carefree Highway*. After the song they got some news, including a breaking story about Bowen Gulch. "Turn it up."

The story was sketchy. Apparently some of the details had been made public about how the deal went down with Western-Pacific. Governor Roy Romer had played a key role. Someone had just leaked a copy of his letter to the Forest Service regional office in which he urged the agency to arrange either a buy-back of the timber in Bowen Gulch, or work out some kind of trade. Public opinion had turned decisively against the timber giant. Badly stung by negative publicity, Western-Pacific had reluctantly bowed to the inevitable, and agreed to the latter option, a trade in timber volume. Leading Republicans had responded with outrage and openly rebuked the governor but to no avail. The segment ended with an interview with an octogenarian grandma, a long-time Granby resident and a well-known pillar of the community. Her great granddaughter had been arrested at the gate with the other protesters. In an aging nasal voice the elderly woman told the reporter, "If I had known what Shelly was doing I would have been there myself. I'm a hiker, you see, an outdoors person. Have been all my life and proud of it too, an' I didn't live 88 years to see those bleep destroy that beautiful place. Not for a chintzy buck..."

"Hmm, I never met Shelly."

A half-hour later they reached the landing.

"Carl won't be thrilled that we missed a day's work."

"He'll get over it."

"We'll make him an offer he can't refuse."

Olsen's skid-jeep was parked outside the tiny camper. He was apparently inside, probably on break, and they heard him as they approached. Carl was talking to somebody in a foreign language.

"Omigod," said Tom.

They banged on the screen door.

"They're back," said Carl. "Come in." He was seated at the small table across from Tallie. The two were carrying on a spirited conversation in Norwegian. The tiny trailer was barely big enough for three let alone four. Somehow Roper ducked in. He stood cramped in the kitchenette.

"Well if it isn't the happy hooligans," the old man said with a harrumph. There was not a trace of the usual vinegar. Tom was startled. Carl was anything but his usual gruff self. He was bare headed, which was also strange. The sweat-stained fedora was sitting on the table next to a teapot and a pair of mugs. The old man was in a jovial mood. Tom had never seen him so expansive. The conversation continued in a lively manner. Tom understood not a word of it. Tallie was apparently explaining something to Carl who exploded with one of his idiosyncratic belly laughs. When it had run its course he pulled out a rag and blew his nose with conviction. Then, he straightened his bifocals and grabbed his fedora. The old man started to slide out of the seat, though the place was packed.

"Now, if you two delinquents will clear me a path," he said, "I've got work to do." Carl cleared his throat. Roper made way. Carl said something to Tallie on the way out but Tom missed it. The old man was straightening his suspenders as he stepped out after them. Carl laid a firm hand on Tom's shoulder and gave him a wink. "Take as much time as you need, son," he said. Then, he shuffled off toward the antiquated crane, still chuckling to himself.

Tom introduced Tallie to Roper, who was looking at her inquiringly. "Dave, this is Tallie McPherson. Tallie, this is Dave Roper."

"Very pleased to meet you."

"We've heard a lot about you," said Roper.

"You have?"

Roper shrugged. “Oh hell yes. You’re all this harelip ever talks about.” He smiled wickedly.

Tom took her by the hand and led her down the road toward his stream side camp. They left his pickup on the landing. Shep started to follow but Roper called him back. Tallie had slung her day-pack over one shoulder. She was also carrying something in a paper bag. He kidded her as they strolled arm-in-arm.

“So what’d you do to him?”

“Do what to whom?”

“To Carl.”

“Didn’t do anything.”

“Yeah you did. I hardly knew him.”

“I was just me. What can I say?”

“You charmed him out of his fedora, nearly out of his suspenders.”

“I did not.”

“What’d you two talk about?”

“Norway, mostly. Carl’s from a small village not far from where one of my cousins lives. South of Oslo, on the coast. He still has family there.”

“I’ve never seen Carl like that.”

“He’s a sweet old man.”

“Sweet? You’re kidding!”

“No.”

“He’s as tough as old shoe leather.”

“He’s the nicest sweetest person.”

“Carl’s as crusty as burnt toast.” There was a pause. He added, “But we like him anyway.”

So it went. They swung their arms. “So, you were able to read the map OK?”

She nodded. “And when I showed it to the bus driver, he knew where to stop. The hike from the highway was longer than I expected though.”

“It’s almost two miles from the turn-off. So, what’s in the bag?”

"Go ahead. Look." She showed him.

"Fresh tomatoes. I love it."

"They're beef hearts. From my aunt's garden."

He looked closer. The tomatoes were plump and ripe as a dream. "Now, we've got all the fixings for a humongous salad. I got a cold-water spring full of watercress; and we got dandelions. They're bitter, you know? But good. Plus other stuff, carrots and wild onions, herbs..."

They walked in silence. When they reached the camp she pulled his arm and kissed him on the mouth. He was already firm.

There was no head-room in the tent. They undressed in the clearing among the lodgepoles. As Tallie slipped her blouse over her head he tickled her. They tussled and fell into the tent, wrestling. The match was a draw. Slowly, their laughter subsided.

There was no shivering. He had prepared a deep nest of blankets, covered by his goose down bag which, unzipped, served as an ultra-lite comforter. Anyway, the day was warm and the evening likewise. The gentlest of summer breezes lulled the tent and the fly. They felt everything around them, the gentle sway of the lodgepoles, Snyder Creek, the sound of water plunging over a beaver dam.

They ended outside of time.

FORTY SEVEN

The sun was more than halfway to its zenith when they finally emerged from the tent, next morning. They made do with a light breakfast of granola, milk and fruit, then lolled about camp in the nude. Tallie was barefoot, in the midst of brushing her silky brown hair when Roper showed up. He had come for a load of spring water. Tallie ducked behind the tent and dressed.

Roper was covered with wood shavings.

"You been working," Tom said as he pulled up his trousers. "I didn't hear your saw."

"I put in a few. I see you've been busy, too."

Tom laughed. "Hey, Dave, we're thinking about climbing Park View."

He and Tallie had discussed it earlier, as they stood in the meadow admiring the enormous mountain that dominated the southwest skyline. Fortunately, she had worn her boots.

"There's an incredible view, up top," Roper said. "You can see most of the state from up there. Windy though."

Tallie joined them. "Morning, Mr. Dave."

"Morning."

"So, *you've* climbed it?"

"Oh sure. Couple of years ago." Roper got out of the truck. He seemed in no hurry to load up his water. He lit a Camel, inhaled deeply, and blew smoke. He stood leaning against the truck.

"It's so amazingly huge," she said.

"It's twelve and a half thousand feet. One big rock pile."

"There's a lot of snow."

Roper turned his head toward the mountain. "Yeah, but the snow is no problem. It's mostly on the north side.

You climb it from the southeast." The cigarette bobbed up and down. He removed the Camel. Smoke scuffed off his tongue. Roper's grin was almost a dare. "It's an easy walk up. You don't need any special gear or anything."

"I have a topo map," said Tom.

"There's an old broke down fire lookout up there. From the 1940s. It's been abandoned for years. Must have been a job building it, 'cause there's no road. There's a huge cairn of stones at the very top, and a survey marker. Carl says he planted it back in '48 or '49. I don't recall. Going on forty years ago."

"What?" Tallie said.

Tom explained. "Carl was a member of the US Geological Survey team that did the geodetic mapping of this part of the state, way back when. After World War II." Roper added, "From the top you can even see the brown smog over Denver. Must be ninety miles away, at least."

"Is Reba back?"

"Yep. Last night."

"Why don't the two of you join us? We'll *all* go."

"Can't. Reba's being ornery this morning. Like to, though."

"She kick you out again?"

Roper shrugged. "I never should have brought her up here in the first place," he said. "She's a city girl. Misses all her citified friends." Roper flicked his ash.

"I love it up here," Tallie said. She had stuffed her hands in her back pockets. Several very fine strands of brown hair had come loose from her pony-tail and hung down in her face.

Tom was so proud of her, then. She was so naturally herself, this waif of a girl. He wanted to show her off to the whole world. He wanted everyone to know that she was his woman. How fetching she was – slender like a new blade of grass. There were so many things to like about her. He liked the slight curve of her hips above her jeans, and her fresh smell, like a wild flower, only better. He liked the graceful way she moved at times. He liked her awkwardness too just as much, and her small breasts. Small was beautiful. He liked the way she held her head up high. She was a proud one, she was, on her good days. And on her bad days he loved her for her amazing courage. He loved her pert manner too. Her musical lilting speech pleased him to no end. Best of all, though, he liked the soulful way she peered out of those beautiful brown eyes of hers. Heck, he even loved those loose strands of hair and how they floated down in her face. Sometimes she blew them out of her eyes, those fine wisps of hair. This time she didn't, maybe she didn't notice them; or maybe she just didn't care. He had half a mind to reach out and brush them back himself, those solitary strands, lighter than air. Not for any particular reason, only because he wanted to and because he liked touching her. She liked it when he touched her, he knew. He did not do it though. He was equally pleased just to admire those loose strands. Even if he could, he would not change a thing, not one strand. He liked her the way she was, wild and free.

"That settles it," Roper said. "I want you to meet Reba."

"We could drop by, later."

"Heck yeah. Come for dinner. I'll cut us some venison steaks."

Tallie's face was a question mark.

"Carl has a large buck curing out in the woods behind his trailer. Suspended from a cross-pole. I'll show you, if you want."

"Why out in the woods?"

"To keep it under wraps."

"Under wraps?"

"Out of sight. The old man poached it."

"Ohhh..."

"What time is it?" said Tom.

Roper peeled back his sleeve. "Already going on noon."

"We're getting a late start."

Roper laughed and blew smoke. He crushed his cigarette out on the tail-gate.

"We'll probably be late getting back. After we do the mountain I'm taking Tallie to the hot springs."

"OK. We'll eat when you get back. *If* you get back. Whenever. Be aware, though, if you don't make it before dark I might have to eat your steaks and mine both." A strange wrinkle moved across Roper's brow. His smile was a taunt. "Strictly out of self-defense you understand."

"Self-defense? Why?" Tallie said.

"Considering the bears."

Her eyes got very wide. "You have bears?"

"Oh sure," Roper said indifferently. "Sometimes they come sniffing around at night."

"You mean, more than one?"

Roper winked. "Last week we had two in one night, didn't we, Tom? My guess is they're attracted to Reba, you know, because of her female scent. Too bad they scare the bejesus out of her."

Tallie's eyes were like globes.

"No sense tempting them with fresh cooked meat," Roper said as he climbed back in his truck. "You two have fun on the mountain."

"Dave, wait a second. I want Tallie to see the bumper sticker on the back of your truck. Tallie, come and check this out." She stepped around to the back of the pickup. The bumper sticker had two parts and read:

UNDER THE DEMOCRATS MAN EXPLOITS MAN

UNDER THE REPUBLICANS ITS JUST THE OPPOSITE

"I like it," she said. "And I totally agree."

Roper gunned the engine, smiling as ever. "Be sure to take along a warm wrap. It can be mighty cold and windy on the mountain."

He left for the spring. When he had loaded the cistern and came back through camp, Tom hitched a ride up to the landing to retrieve his pickup. As he was getting out, Roper said, "Why don't you take Shep along. Do him good. He needs a workout."

"Sure."

"Shep, you go with Tom." The dog leaped down out of Roper's truck, then up again into the back of Tom's, wagging his tail.

"See you."

"Later, bud."

They made ready, packed a lunch, sandwiches, apples, cheese, and raisins, chocolate bars, even a thermos of freshly brewed coffee. They also packed a water bottle, sweaters and windbreakers. But they never made it to Park View Mountain. They never made it out of camp. They ended up, same as before, shacked up in the little tent where they spent a good part of the afternoon in their private reality. Shep was more than content to snooze in the clearing.

FORTY EIGHT

Eventually they rose and dressed. Within the hour they were cruising north toward Walden. Slowly the enormous mass of Park View receded in the rear view. Shep rode in the back of the truck, nose forward, the wind preening his fur.

"What a fine dog."

"And smart too. How's your head?"

"Steady." She was having another pain-free day. The second in a row.

"Marvelous." They rode with the windows down, the draft in their faces. The July day was near perfect, dry and warm. There was hardly a cloud in the sky.

The country was wide, and the scenery bountiful with colors and vistas on every side. They passed occasional ranches and a few cabins. The roadside was a blush of purple asters, blue bells, and sweet lupine. The ditches brimmed with snowmelt, the water bright in the sunlight. Patchwork fields lay soaking on both sides of the highway. In places the meadows were like a rainbow. High grass was in full tassel along the revetments.

The road hardly wavered from its northwest heading. Tom cruised with a lazy finger on the wheel.

They rode in silence. They had become telepathic. He pointed to a hanging snow-field on a high ridge, west of the highway. It was breathtaking. She smiled and nodded. No need to speak.

The country changed. Now, they motored through the big-belted sage of North Park's rolling mid-section. Occasional herds of cattle could be seen grazing the distant sage-flats. A few pronghorns dotted the sparse hills.

Just south of Walden he turned west on the highway to Rabbit Ears Pass. He followed it for a mile or so, then turned again onto an unimproved county road with a due west heading. Up ahead, the gravel road vanished into distance. Before them rose the Park Range, crowned by Mt. Zirkle.

"It's hard to believe this goes someplace."

"I know. But it does. You'll see."

"So, where are the hot springs?"

"At the foot of those high peaks. About seven miles ahead."

"You've been to them?"

"Sure. We learned about them from Carl. The best part is that they were never developed commercially. The springs supposedly belong to the Walden Odd Fellows Club."

"What a strange name."

"Yes, how very *odd.*"

"Smart ass."

"They're kind of like the Lions, I guess, or the Elks. Carl says the springs were man-made."

"How could hot springs be man-made?"

"Back in the fifties some wildcat driller hit mineral water instead of oil. Hot mineral water; and they've been flowing ever since, except for a short time in the 1960s when they ran dry. According to Carl, a bunch of hippies had moved in and set up camp around the springs. Nobody cared much, at first, but eventually the hippies became a nuisance because they wouldn't leave. The Odd Fellows sent out the local sheriff and a posse to chase them off. But every time they evicted the squatters, the hippies would leave for awhile, then come back when the coast was clear. This went on until the Odd Fellows finally got fed up."

"What did they do?"

"They sent out a bulldoze operator."

"To what?"

"Bury them. They plowed them under."

"That's so sad."

"It didn't work, though. There was too much pressure from underground. After a few months the hot springs opened up again on their own. And I guess they've been running free, ever since, at a constant 106 degrees, summer and winter. It's why they're so good. They're plenty hot. For some reason, though, the hippies never came back."

"Oh, yes they did."

There was mischief in her eyes. "Oh, I get it. I am you, and you are me, and we are all together."

"You are the Walrus."

"I am the Walrus."

"Ku-ku-kuchub!"

They passed several miles in silence. The road had narrowed. They motored around a bend. "Look," he finally said. "You can see 'em. Across that pasture."

"The hot springs? Where? Show me."

"See the steam rising along the far edge of that meadow. Over there. If you look close."

"Yah! Yah! I see the steam!"

"I wouldn't be surprised if these are the best hot springs in the Rockies."

"Why are they so good?"

"Well, they are not on any map. Usually, there's nobody around."

They slowed and crossed a cattle guard, then, came to a fork. He took the right branch. Turning again, he made another right and followed a rutted jeep trail along an old fence line. The track eventually played out at the edge of a cow pasture. He parked beside an irrigation ditch, full to the brim with fast-moving water. There were no other cars. They had the springs all to themselves. He turned off the engine. The only sound was the gentle breeze whispering in the prairie grass, and water gurgling. A meadowlark warbled in a nearby willow, a flute-like melody.

"Let's do it!"

"Yippee!"

Tallie grabbed the towels and they spilled out. Shep was gone in an instant, sniffing here and there, this and that, checking out the meadow, that is, after he anointed the nearest fence post.

To reach the springs they had to negotiate a barbed-wire fence; but it was not a serious obstacle. Someone had thoughtfully constructed wooden stairs, which allowed them to step up and over the wires.

By now, Shep was far out in the meadow chasing his nose.

It was a pleasure to stroll the last hundred yards to the steaming springs along the irrigation ditch bordering the meadow. By foot seemed the proper way to arrive at such a place. The springs were five hundred feet higher than the wide floor of North Park, and the view across the valley to the distant Medicine Bow and Never Summer ranges was spectacular. They crossed a small footbridge, just two planks thrown over the ditch, and arrived at the main tank. It fed a series of lesser pools. The temperature of each decreased in stages, as the hot mineral water flowed from one pool down to the next. There was thus a temperature to suit every taste. The last tank was shallow and quite cool. The grass was deep and green where it drained out and the spring finally piddled away.

Without a word Tom leaped out of his clothes into the main tank. Tallie followed. The next moment they were up to their necks, arms floating free, hot mineral water bubbling up from deep underground, tickling their toes and massaging their bare skins.

"I feel ten pounds lighter," Tom said. She had managed to sneak around behind him and pushed his head under. He got a good ducking but spun away and came up splashing. A furious water fight ensued. However, it soon ended in the middle of the pool where they came together. Another draw. He was hard before she touched him. He placed his hands firmly on her shoulders as she took him in, shuddering slightly. Her pony-tail was soaked, flat against her flushed skin.

Their unhurried coupling in the buoyant water was amazing. The mineral spring served to heighten their senses and, together they floated near to eternity. But the heat was cumulative and in the end it proved too much. They separated before they were done, abandoned the spring and finished casually in the cool grass beside the tank.

After toweling off, they dressed in silence, then, lazed about in the westering sun, amused by the great vole hunt unfolding out in the meadow. Shep the mighty hunter would sometimes double back with his nose down. Each time he sniffed out his quarry he would leap straight up, two feet in the air, at least; all four paws up at the same instant. He pounced the moment he came down.

"Simply incredible."

"It's the coyote in him."

The dance number was comical but effective. Shep bagged several snacks as they watched.

Slowly the late afternoon sun dropped from sight behind the near ridge. The springs passed into shadow. As the day retreated across the meadow the line of shadow perceptibly lengthened across the valley floor. The paradox of the fading sun was that as it sank lower in the West the snow-capped mountains in the East were illumined all the more brightly, until all colors merged in a brilliant gold.

"What a sight."

"Golly, it's nice country."

There was a long silence. "Tallie..."

"Yes?"

"Did you ever feel like..." He paused, then started again. "You know, sometimes I think all I want out of life is to go up into those mountains. And maybe never come back. Does that sound nuts to you?"

"Not to me."

"Did you ever feel that way?"

"It would be a lot of fun. Tom, you know what?"

"What?"

"Earlier this afternoon, that's what I was thinking. Remember, I told you I used to spend summers at my aunt Mary's ranch. When I was very young. What I didn't tell you is, they used to take me up into those mountains. But it was so many years ago. I had almost forgotten. I was thinking what fun it would be to do it, again. You and I. To go up there...and explore. What a name. Never Summer. It's easy to see why they call them that."

"Most years, the snow never melts. Last year, there was almost as much snow on the high peaks in September as in June. Damn near. There are lakes and springs up there too if you know where to look. And a few small glaciers"

"I *really* want to see Bowen Gulch. Could we go there?"

"Sure. We could follow the Illinois River, all the way in. You know the little creek by my camp?"

"Snyder Creek?"

"Yes. Well, two miles downstream it flows in to the Illinois."

"See that notch on the horizon." He pointed across the valley. "Right there. That's where the Illinois comes out of the Never Summers." He showed her. "Follow my arm." She did.

"You mean...that little dip?"

"Yes. We could hike downstream to the Illinois, then make a right turn and follow it back up into the mountains. All the way in. Bowen Pass is at the head of that valley. Once you get above timberline you can tromp for miles through wild meadows and tundra and be as far from civilization as a person can get. You can see about everything that's worth seeing. Carl says the view from up there is incredible. Then we could drop down the other side into Bowen Gulch. It would be easy. The pass is on the divide. You'll love the shaggy old spruces. They blew me away. The biggest and oldest trees in Colorado...."

They made it back to Roper's in time for dinner.

Later that night, he was awakened by a freshening wind in the treetops. He rolled over. The blow was tugging at the guy lines. He could feel the pines swinging and swaying. Tallie slept peacefully beside him, her quiet breath rhythmic, unlabored, her dreams undisturbed. For a moment he studied the profile of her face. He felt restless like the night. Quietly he slipped out from under the goose down, pulled on his pants and a light sweater, slid into his shoes, grabbed his flashlight and went out.

A crescent moon was up. There was more than enough pale light to see by, and to move about camp without tripping or bumping into things. Slowly, deliberately, he made his way through the lodgepoles to the edge of the meadow. He sat on the grassy slope listening to the wind. How he loved these unsettled nights. Frogs were sounding off down by the creek. "Galumph! GALUMPH! Galumph!"

A few clouds raced before the moon. Something stirred in him. He felt the start of something. What? He had no idea. Maybe another poem. Whatever, it was still formless. He moved to the temporary table near the fire pit. He had fashioned it by lashing together some poles and old boards. It was crude but served the purpose of a "camp table" well enough. He decided to forego the lantern. He didn't want the hissing to distract him. He wanted to hear the night around him, especially the wind. He sat down, snapped on his flashlight, and laid it sideways.

That will do.

He could work nicely within its parabolic beam. He felt now that the thing inside him was a poem trying to be born.

He was pregnant and his labor was about to begin. He picked up the pen, but set it down again and cracked his knuckles. No hurry. He could wait. Whatever was coming would happen at its own speed, in its own time. He picked up the pen again and doodled on the yellow pad. Then he wrote the first thing that came into his head. Once the words started they flowed out. They were attached to feelings. He worked fast. The first verse came out like a fetus, perfectly formed. No changes were needed. Here is what he wrote:

There's nothing like a skinny
girl with freckles on her nose,
a barefoot girl with knobby
knees and grass between her toes.

He snapped off the light and waited. He sensed another verse coming. It would happen when it was good and ready. He chewed on the pen. The words were in the folds of the night, on the wind. All he had to do was wait and allow the inner voice to speak through him. A moment later he snapped on the flashlight and wrote:

There's nothing that a naughty
girl will hesitate to do.
A willing girl with sassy
eyes can make your dream come true.

Good and spunky! He liked it. Should he try for a third verse? Yes? No? He wasn't sure if more would come. Maybe he was done. He was thankful for what he had received already. No point in pushing his luck. Maybe it would come and maybe it would not. He turned off the light, got up, and prowled about in the dark. Suddenly, a strong wind came up. It was trying to snatch the treetops away.

Yes!

That was a part of it. The rest was coming. Now, he had it. He returned to the table and clicked on his light and wrote:

There's nothing like a pony
tail to take your breath away,
a ponytail, a pretty
girl! And nothing left to say...

Later, she awakened, her head a void. No pain. He was beside her, sleeping soundly, sawing logs, snoring loudly. She listened to his rhythmic breathing. She smiled as she recalled the story Roper had told at dinner, no doubt, for her benefit. She liked his friend. Something he'd heard from Carl Olsen, but Roper didn't think it originated with Carl, as it wasn't his style. Probably just something he had picked up over the years on one of his hunting expeditions. A tale of three hunters and a guide, but only two tents. So, they had to double up. So, well after the first night one of the men turned out grumpy complaining that he had not slept a wink. Why not? They asked him. Because, he explained, the man beside him had snored all night, *very* loud, and it kept him awake. The second night, the hunter who had not slept well insisted on trading partners. So they switched and sure enough, next morning the new mate showed up at breakfast with the same complaint. He'd passed a near sleepless night because of the loud snoring. On the third night they switched again. This time, it was the hunting guide's turn to sleep with the snoring wonder. But in the morning when they asked, he told them he'd slept just fine, like a pile of rocks in fact. Never slept better in my life. But didn't the loud snoring bother you, didn't it keep you up? No, the guy didn't snore at all, he told them. But how is that? The cagey guide explained that before turning in he gave the man a kiss on the forehead, then, dropped right off to sleep. He guessed the fellow never slept because once in the wee hours he woke up and noticed he was wide awake, watching him.

Tom was snoring as ever. She rolled over and closed her eyes, smiling. Within seconds she joined him in dreamland.

FORTY NINE

In the morning, he drove her into Walden. She planned to catch the bus from Granby to Denver where her cousin would pick her up. They left in time to have a leisurely breakfast in town. The bus was not due until 10:30 A.M.

The drive was uneventful. They hardly spoke. By now they were in their own quiet space. He held her hand part of the way.

As he motored into Walden Sunday church bells started chiming from one end of town to the other. The place had more than its share of churches, each with its own set of differently scaled bells. The result was a medley. Tom wondered half-seriously if the town was serendipitously announcing their arrival. The answer came in the deserted streets. Walden was a near ghost town.

"Where is everybody?"

"Maybe in church?"

"Everyone? The *whole* town?"

As they cruised up main street tumbleweeds rolled across the dusty avenue. It was broad but only three blocks long, a nostalgic ruin of a place straight out of *The Last Picture Show.*

He parked by the First National Bank, across from the Elkhorn Bar & Grille. The chiming continued as they strolled across the wide street and stepped up on the high curb. The cafe's doorbells added to the tintinnabulation as they entered the restaurant.

The Elkhorn was actually two establishments in one. The saloon side was closed and dark. But the restaurant half had just opened for business. They were the first customers of the day. Tallie picked a booth by the front window and set down her satchel. Tom dropped his too, then poked his head through the swinging double doors into the dark lounge. Large trophy-heads glowered back at him from the shadows: bull elk, a large buck or two, several moose with huge racks, cougar, bear, and bobcat.

The cook appeared in a clean white apron with a steaming pot of coffee and two menus under his arm. The waitress apparently was late for work. Tom ordered his usual, a three-egg omelet with hash browns, whole-wheat toast, and jam. Tallie wanted a stack of buckwheat cakes. While they waited they slurped their coffee.

The place had a 1950s feel. The tables were set with red-and-white-checkered tablecloths. The red vinyl upholstery in the booths was cracked and worn from use. The plaster walls were paneled halfway up in the old style, with custom moldings. Above the wainscoting the cream-colored walls were covered with yellowing photos of old-time cars, logging trucks, railroad engines, and other heavy machinery, most of which was unknown to Tom. He guessed it was mining equipment. In some of the pictures men in derby hats proudly posed for the camera outside their business establishments; Walden in a bygone era.

He fiddled with his fork. Their eyes met. "What?"

Reaching down, he removed two books and some papers from his satchel. He pushed the books across to her. "I borrowed these from Mary. Would you return them?"

"Oh sure."

"The other one I'm not finished with yet. I'd appreciate it if you'd tell her thank you for me."

"I'm sure she's happy to loan you books anytime. What are you reading?" She studied the top book. "Were they good?"

"Damn good."

"This looks deep."

"It is. Emerson was right about a lot of stuff."

"So when do you read?"

"Evenings. Y'know, I like the physicality of the work. I love the hell out of running a chainsaw. I really do. The feeling of the saw eating wood is an adrenaline rush. I can't explain it. A full day of it leaves you feeling beat, but I don't mind. I like knowing, feeling that I did something real. With my hands. Only, there's no intellectual challenge to it. So, I read at night."

"You use a lantern? Like in Florida?"

"Yeah."

She set one book aside and opened the other one. "Hey, this one's about the mushroom." She stared at the cover, then opened it. "I know this. I've seen these at the ranch."

"That's what Mary said."

"I didn't get to try it. I want to."

He had already told her about the caprice of Mother Nature. "You may have to wait awhile. Maybe ten to twenty years."

"I won't be around in twenty years."

"It doesn't matter. You and I don't need mushrooms."

"What a sweet thing to say."

He set the poems on the table before her. Almost at once he regretted doing it. He had decided they were pretty bad. He felt good about them when he wrote them but when he reread them they seemed inadequate. Perhaps because they fell so far short of his true feelings, what he was trying to say. The poems barely scratched the surface. For a long time, ever since Florida, he had wanted to tell her how he felt about her. There was so much, but he felt he always botched it somehow. He generally stumbled over his feelings and had a hard time expressing himself. From one moment to the next he was at a loss to say how he felt. Why was it so hard?

"I know they're not much."

Tallie looked up. She had been reading them. "Oh no. They are wonderful. I mean it. You're a writer."

"Tallie, I..."

She saw his discomfiture and reaching across, put a finger on his lips. "Shhhh. Hey you. I love them."

"I..." He started again but gave it up.

She read the poems again. After a long pause she looked up and gave him a searching look.

"What?"

"Oh nothing."

But he knew she wanted to say or ask him something. "Go ahead."

"Do I really have ... knobby knees?" Her brow was wrinkled, her face almost a grimace.

Laughing, he squeezed her hand. "Yes, as a matter of fact, you do. And what delicious knees they are!" He kissed his fingers and blew the kiss to the winds. "Ooo-la-la! How I love zeez kneez!" Her face cracked and she laughed.

A young woman strode into the restaurant. She was obviously in a hurry and disappeared into the kitchen.

"The waitress."

"Hah. Late."

"For a very important date."

The girl reappeared tying her apron strings behind her back. Now, the doorbells jingled. Customers began trooping in, talking and laughing, a jostling crowd of people.

"One of the churches just let out," Tallie said, her eyes full of mischief.

"You naughty thing."

"One of the ministers gives verrrry short sermons."

"My kind of guy." They laughed.

More customers came in, one group after another. They were locals, working men and their families, moms with young babies, and kids large and small, boys with cowboy hats and matching boots and little girls with pigtails tied with pink ribbons. There were old folks too. All of them were dressed up, in their Sunday finest. Within minutes every table and booth was occupied. The cafe became noisy with conversation and the sounds of hungry people. The waitress brought their food, then returned to freshen their coffee.

"Good thing we came early."

"Don't know about you but I'm famished."

"Me too."

"That is one pile of flapjacks."

"If I'd known I'd have ordered a short stack."

They gabbed while they ate, about one thing and another. Small talk. Nothing of any importance. Then she blurted out, "Oh hey, I know how we'll do it."

"Do what?"

"The trip to Bowen Gulch."

"How?"

"Last night, it came to me. We'll go on horseback. You can ride Luther."

"I love it."

"Why walk when you can ride?"

"Why didn't I think of that?"

"We'll use a pack horse to carry our gear."

"Yes, but do you think your aunt will let us use her horses? She is really attached to them."

"I'll ask when I get back but I'm sure. Almost positive, in fact. But I'll need your help. Can you pull a horse trailer?"

"Well, my truck's got a ball on the bumper."

"I do have a license believe it or not; but I never pulled my aunt's trailer."

"They're tricky, especially backing and turning."

"You know how?"

He nodded. "Yes, but I don't think my four cylinder pickup can handle a horse trailer."

"No?" She pouted.

"Too much weight. But don't worry. Roper will let me use his truck. He drives a V-8, with four-wheel drive. Loads of power."

"How do you know he will let you drive it?"

"He will. Don't worry. He's a friend." Tom got up. "Wait," he said. "I'll be back in a sec." He went over to the counter. Tallie continued eating. When he returned he laid a Sunday paper on the table. "You reminded me. Here. Look at this."

"What?"

"Check out the headline."

The front page read: SPECIAL WILDERNESS BILL PASSES HOUSE. The sub header was in smaller print: BOWEN GULCH SET ASIDE.

"They finally got the bill out of committee."

She was reading. "So that's it."

"Yes, it feels good to win one for a change."

"I can hardly wait."

"Bowen Gulch here we come."

"We'll plan it."

For awhile they sat in silence. Tom was done but Tallie was still eating.

"The waitress arrived with a pot. "More coffee?"

Tom put a hand over his cup. "None for me, thanks."

"No," Tallie said. She smiled at the waitress, then went back to mopping up her syrup.

Brakes squealed. A silver Greyhound pulled up along the high curb. There was the loud release of brakes. Compressed Air.

"That's my ride. Yikes. No way I'll finish." Even so, she kept at it.

"That's the biggest pile of cakes I ever saw."

"Maybe I should get a doggy bag?"

"Right. Take it with." He wiped his chin and got up with check in hand. He went over to the antique cash register. He returned with a toothpick in his mouth, then, sat and quietly watched her eat. She was still going for it.

Finally she set her fork down and pushed the plate away. "Enough." She dropped her napkin on the table. "I'll be right back." She disappeared into the ladies' room.

While he waited he inserted the books into her daypack and buckled it up. When she returned they grabbed their stuff and went out.

The Greyhound was waiting to take on passengers. Among its various civic functions, the versatile Elkhorn served as the Walden bus depot. The bus door and luggage compartment were open. The powerful diesel engine was humming. The driver nodded a friendly "Good morning."

Tallie handed him her ticket. He punched it and passed it back, then, stowed her gear. She was the only departing customer. The bus was nearly empty. There were only two other riders. The driver glanced at his watch. "We're running a few minutes late," he said. "But I need a cup of java. Won't be a minute." Loose change jingled in his pants as he strode across the sidewalk and into the cafe. The smell of cologne lingered after him.

They waited curbside. She was happy for the delay, but Tom hated goodbyes and stood flatfooted with his hands in his pockets. He did not know what to say. She studied him as she let down her pony-tail. She put the tie-band in her teeth and shook her head from side to side to loosen her long silky hair. She gathered it up again and slipped the band back in place. She stepped up and kissed him on the cheek, one messy smack. Playfully she combed her hand through his blonde hair.

"What a mop."

"I'm always putting you on a bus to somewhere."

She cut him off. "But this time it won't be six months."

"You promise?"

"I don't have six months," she blurted out.

"You..."

"Oh, Tom. Will you call me tomorrow at my cousin's? In the evening."

"Bet on it." He pressed her hand against his chest. Tears began streaming down her face.

Oh shit, she's crying.

"Please don't."

"I'm not," she said, wiping one eye with the back of her other wrist. "It's just that I'm ... so happy!" The next moment she was giggling, pleased as a kitten. But the tears continued to flow.

"Christ." He wiped them away with his hand, then, slipped his arms around her and kissed her on the button. The driver returned.

"Hate to break up the party, kids. We got to hit the road. All aboard!"

They lingered another long moment. She gave his hand a squeeze. Then, she climbed into the bus. He watched her make her way down the aisle to the seat by where he stood. With both hands she unsnapped the window and slid it open. She rested her chin on her arms.

"I have to go."

"I know. But, next time..."

Her face brightened. "Can I stay a week?"

"Stay as long as you want."

"Yay! A week in the Never Summers!"

"We could see a lot of country in a week."

"You'll..." But she was drowned out by the diesel engines. The Greyhound gahoooomed and a nasty cloud of soot poured out of the tail pipe.

She waved.

He waved back. The Greyhound started to roll. He watched her move to the rear of the bus, her ponytail bobbing up and down. A moment later, her face appeared in the back window. One last wave and a big smile.

From the Elkhorn it was one short block to the edge of town where Main Street ended in a sage pasture. Open country stretched without a break to the horizon. For three miles north the highway to Cowdrey dropped off the bluff and ran true as an arrow, parallel with the power line poles.

From the high curb he watched the bus grow small, until it was only a dot. The silver speck lingered on the crest of a rise, then was gone.

He fumbled through the loose change in his pocket until his hand closed around the keys. For a moment he looked down at his boots. Then, he turned toward his pickup.

Acknowledgments

Many individuals assisted in one way or another in the birthing of Never Summer. My sincerest thanks to all of them, including conservation biologist Reed Noss, Bruce Means of the Coastal Plains Institute and Land Conservancy, Laurie MacDonald of Defenders of Wildlife, Robert VanNatta, for his invaluable expertise about logging equipment, and the late Don Hudson. The final incarnation of the book would not have been possible without videographer Gary Huber, forest activist Rocky Smith and professor Marty Walter, all of whom provided historical details about the campaign to save Bowen Gulch. Forest Service officials who helped include Rick Cables, Karen Roth and Jeff Underhill. Also, a BIG thank you to the following who read and/or edited the manuscript: Jean Weininger, Jeanie Shaterian, Joyce Merwin, Scott Dickerson, Ed Atkin, Victoria Tenbrink, and Kelly Ray. Their feedback was indispensable. Photographer Chris Hanson gets credit for the beautiful photo gracing the cover. Finally, a big thank you to the late Mike McCann, who kept after me until he finally persuaded me to finish the book.